D0192080

New National Qualifications

Practice Papers for SQA Exams

Higher

Mathematics

© 2014 Leckie & Leckie Ltd

001/10122014

10 9 8 7 6 5 4 3 2 1

ISBN 9780007590919

Published by
Leckie & Leckie Ltd
An imprint of HarperCollins*Publishers*
Westerhill Road, Bishopbriggs, Glasgow, G64 2QT
T: 0844 576 8126   F: 0844 576 8131
leckieandleckie@harpercollins.co.uk   www.leckieandleckie.co.uk

Publisher: Peter Dennis
Project Manager: Craig Balfour

Special thanks to
Ink Tank (cover design)
QBS (layout and illustration)
Felicity Kendall (proofreading)
Peter Lindsay (proofreading)

A CIP Catalogue record for this book is available from the British Library.

Acknowledgements
Whilst every effort has been made to trace the copyright holders, in cases where this has been unsuccessful, or if any have inadvertently been overlooked, the Publishers would gladly receive any information enabling them to rectify any error or omission at the first opportunity.

Introduction

Layout of the Book

This book contains practice exam papers, which mirror the actual SQA exam as much as possible. The layout, paper colour and question level are all similar to the actual exam that you will sit, so that you are familiar with what the exam paper will look like.

The solutions section is at the back of the book. The full worked solution is given to each question so that you can see how the right answer has been arrived at. The solutions are accompanied by a commentary which includes further explanations and advice. There is also an indication of how the marks are allocated and, where relevant, what the examiners will be looking for. Reference is made at times to the relevant sections in Leckie & Leckie's book *Higher Maths Success Guide*.

Revision advice is provided in this introductory section of the book, so please read on!

How to use this Book

The Practice Papers can be used in two main ways:

1. You can complete an entire practice paper as preparation for the final exam. If you would like to use the book in this way, you can either complete the practice paper under exam style conditions by setting yourself a time for each paper, and answering it as well as possible without using any references or notes. Alternatively, you can answer the practice paper questions as a revision exercise, using your notes to produce a model answer. Your teacher may mark these for you.

2. You can use the Topic Index at the front of this book to find all the questions within the book that deal with a specific topic. This allows you to focus specifically on areas that you particularly want to revise or, if you are mid-way through your course, it lets you practise answering exam-style questions for just those topics that you have studied.

Revision Advice

Work out a revision timetable for each week's work in advance – remember to cover all of your subjects and to leave time for homework and breaks. For example:

Day	6pm–6.45pm	7pm–8pm	8.15pm–9pm	9.15pm–10pm
Monday	Homework	Homework	English Revision	Chemistry Revision
Tuesday	Maths Revision	Physics Revision	Homework	Free
Wednesday	Geography Revision	Modern Studies Revision	English Revision	French Revision
Thursday	Homework	Maths Revision	Chemistry Revision	Free
Friday	Geography Revision	French Revision	Free	Free
Saturday	Free	Free	Free	Free
Sunday	Modern Studies Revision	Maths Revision	Modern Studies	Homework

Make sure that you have at least one evening free a week to relax, socialise and re-charge your batteries. It also gives your brain a chance to process the information that you have been feeding it all week.

Arrange your study time into one hour or 30 minute sessions, with a break between sessions, e.g. 6pm – 7pm, 7.15pm–7.45pm, 8pm–9pm. Try to start studying as early as possible in the evening when your brain is still alert and be aware that the longer you put off starting, the harder it will be to start!

Study a different subject in each session, except for the day before an exam.

Do something different during your breaks between study sessions – have a cup of tea, or listen to some music. Don't let your 15 minutes expanded into 20 or 25 minutes though!

Have your class notes and any textbooks available for your revision to hand, as well as plenty of blank paper, a pen, etc. You should take note of any topic area that you are having particular difficulty with, as and when the difficulty arises. Revisit that question later having revised that topic area, by attempting some further questions from the exercises in your textbook or referring to the *Higher Maths Success Guide*.

Revising for a Maths Exam is different from revising for some of your other subjects. Revision is only effective if you are trying to solve problems. You may like to make a list of 'Key Questions' with the dates of your various attempts (successful or not!). These should be questions that you have had real difficulty with.

Key Question	1st Attempt		2nd Attempt		3rd Attempt	
Textbook P56 Q3a	18/2/15	✗	21/2/15	✔	28/2/15	✔
Practice Exam A Paper 1 Q5	25/2/15	✗	28/2/15	✗	3/3/15	
2014 SQA Paper, Paper 2 Q3	27/2/15	✗	2/3/15			

The method for working this list is as follows:

1. Any attempt at a question should be dated.

2. A tick or cross should be entered to mark the success or failure of each attempt.

3. A date for your next attempt at that question should be entered. For an unsuccessful attempt: 3 days later; for a successful attempt: 1 week later

4. After two successful attempts remove that question from the list (you can assume the question has been learnt!)

Using 'The List' method for revising for your Maths Exam ensures that your revision is focused on the difficulties you have had and that you are actively trying to overcome these difficulties.

Finally, forget or ignore all or some of the advice in this section if you are happy with your present way of studying. Everyone revises differently, so find a way that works for you!

Transfer your Knowledge

As well as using your class notes and textbooks to revise, these practice papers will also be a useful revision tool as they will help you to get used to answering exam style questions. You may find as you work through the questions that you find an example that you haven't come across before. Don't worry! There may be several reasons for this. You may have come across a question on a topic that you have not yet covered in class. Check with your teacher to find out if this is the case. Or it may be the case that the wording or the context of the question is unfamiliar. This is often the case with reasoning questions in the Maths Exam. Once you have familiarised yourself with the worked solutions, in most cases you will find that the question is using mathematical techniques with which you are familiar. In either case you should revisit that question later to check that you can successfully solve it.

Trigger Words

In the practice papers and in the exam itself, a number of 'trigger words' will be used in the questions. These trigger words should help you identify a process or a technique that is expected in your solution to that part of the question. If you familiarise yourself with these trigger words, it will help you to structure your solutions more effectively.

Trigger Word	Meaning/Explanation
Evaluate	Carry out a calculation to give an answer that is a value.
Hence	You must use the result of the previous part of the question to complete your solution. No marks will be given if you use an alternative method that does not use the previous answer. (unless "or otherwise" appears after "hence")
Determine	Usually you should find a numerical value or values
Expand	This means different things in different contexts: 　　Algebraic expressions: remove brackets 　　Trig examples: use the addition formulae sin(A±B) or cos(A±B)
Show that	You should include every step of your working
Simplify	This means different things in different contexts: 　　Surds: reduce the number under the root sign to the smallest possible by removing square factors. 　　Fractions: one fraction, cancelled down, is expected. 　　Algebraic expressions: get rid of brackets and gather all like terms together.
Express	This usually means that you are required to rewrite what you are given in a particular form.
Algebraically	The method you use must involve algebra, e.g. you must solve an equation or simplify an algebraic equation. It is usually stated to avoid trial-and-improvement methods or reading answers from your calculator.
Justify your answer	This is a request for you to indicate clearly your reasoning. Will the examiner know how your answer was obtained?
Show all your working	Marks will be allocated for the individual steps in your working. Steps missed out may lose you marks.

In the Exam

Watch your time and pace yourself carefully. Some questions you will find harder than others. Try not to get stuck on one question as you may later run out of time. Rather return to a difficult question later. Remember also that if you have spare time towards the end of your exam, use this time to check through your solutions. Often mistakes are discovered in this checking process and can be corrected.

Become familiar with the exam instructions. The practice papers in this book have the exam instructions at the front of each exam. Also remember that there is a formuae list to consult. You will find this at the front of your exam paper. However, even though these formulae are given to you, it is important that you learn them so that they are familiar to you. If you are continuing with Mathematics next session it will be assumed that these formulae are known in next year's exam!

Read the question thoroughly before you begin to answer it – make sure you know exactly what the question is asking you to do. If the question is in sections (e.g. 15a, 15b, 15c, etc), then it is often the case that answers obtained in the earlier sections are used in the later sections of that question.

When you have completed your solution, read it over again. Is your reasoning clear? Will the examiner understand how you arrived at your answer? If in doubt then fill in more details.

If you change your mind or think that your solution is wrong, don't score it out unless you have another solution to replace it with. Solutions that are not correct can often gain some of the marks available. Do not miss working out. Showing step-by-step working will help you gain some marks even if there is a mistake in the working.

Use these resources constructively by reworking questions later that you found difficult or impossible first time round. Remember: success in a Maths exam will only come from actively trying to solve lots of questions and only consulting notes when you are stuck. Reading notes alone is not a good way to revise for your Maths exam. Always be active, always solve problems.

Good luck!

Topic Index

Topic	A Paper 1	A Paper 2	B Paper 1	B Paper 2	C Paper 1	C Paper 2
Unit 1 Expressions and functions						
Functions and graphs	4		2,12a	8a,b	2,3	
Exponential and logarithmic functions	9	6	3,7,12a		11	6
Trigonometric functions	1,11a		5	6a,b	10,12a,12b	2a
Vectors	5	1c,4	4,6	5	5	3
Unit 2 Relationships and calculus						
Polynomials	2,10		8,12b		5	8
Trigonometric equations	11b,12	8		4,6c	12c	1,2b
Differentiation	8	1a,2	11	1,6c	4,9	5,8
Integration	3		1,9		8	11
Unit 3 Applications						
The straight line	6	3		3	6	
The circle	7	1b		2		7
Sequences and recurrence relations		5	10		7	
Applications of differentiation and integration		7		7,8c,8d		9,10

This table shows how the various Unit 1, Unit 2 and Unit 3 topics are distributed throughout the Practice Papers. You may find this information helpful if you particularly wish to focus on the revision of one particular topic.

Practice Paper A

Higher Maths Practice Papers

Practice Papers for SQA Exams

Fill in these boxes and read what is printed below.

Full name of centre

Town

Forename(s)

Surname

Total marks — 60

Attempt ALL questions.

You may NOT use a calculator.

Full credit will be given only to solutions which contain appropriate working.

State the units for your answer where appropriate.

Leckie×Leckie

Scotland's leading educational publishers

FORMULAE LIST

Circle:

The equation $x^2 + y^2 + 2gx + 2fy + c = 0$ represents a circle centre $(-g, -f)$ and radius $\sqrt{g^2 + f^2 - c}$

The equation $(x - a)^2 + (y - b)^2 = r^2$ represents a circle centre (a, b) and radius r

Scalar Product: $\quad a.b = |a||b| \cos \theta$, where θ is the angle between a and b

or $\quad a.b = a_1 b_1 + a_2 b_2 + a_3 b_3$ where $a = \begin{pmatrix} a_1 \\ a_2 \\ a_3 \end{pmatrix}$ and $b = \begin{pmatrix} b_1 \\ b_2 \\ b_3 \end{pmatrix}$

Trigonometric formulae:

$$\sin (A \pm B) = \sin A \cos B \pm \cos A \sin B$$

$$\cos (A \pm B) = \cos A \cos B \mp \sin A \sin B$$

$$\sin 2A = 2\sin A \cos A$$

$$\cos 2A = \cos^2 A - \sin^2 A$$

$$= 2\cos^2 A - 1$$

$$= 1 - 2\sin^2 A$$

Table of standard derivatives:

$f(x)$	$f'(x)$
$\sin ax$	$a \cos ax$
$\cos ax$	$-a \sin ax$

Table of standard integrals:

$f(x)$	$\int f(x)\, dx$
$\sin ax$	$-\dfrac{1}{a} \cos ax + C$
$\cos ax$	$\dfrac{1}{a} \sin ax + C$

1. (a) Expand $\cos(x - 30)°$

1

(b) Hence, or otherwise, find the exact value of $\cos 15°$

3

2. Find the range of values of k for which $(2k + 3)x^2 - kx - 1 = 0$ has real roots where $k \in \mathbb{R}$

4

3. Find $\int \dfrac{2x^5 - 3}{3x^4}\, dx$

4

4. Two functions are defined by $f(x) = \frac{1}{2}x + 1$ and $g(x) = 3x - \frac{1}{2}$ where $x \in \mathbb{R}$

(a) Find a formula for $h(x)$ where $h(x) = g(f(x))$

2

(b) $h^{-1}(x) = px + q$ where p and q are constants. Find the values of p and q.

3

5. Six points are located in space.

The following facts are known:

- A has coordinates $(-1, 2, 3)$
- $\overrightarrow{AB} = 4\mathbf{i} - 2\mathbf{k}$
- $\overrightarrow{BC} = \overrightarrow{AB} + 2(\mathbf{i} + \mathbf{j} + \mathbf{k})$
- $\overrightarrow{AB} = -2\overrightarrow{CD}$ and $\overrightarrow{BC} = 2\overrightarrow{DE}$

Find the coordinates of the point E.

B

$A(-1, 2, 3)$

C

D

E

4

6. The line l_1 makes an angle of $\frac{\pi}{3}$ with the positive direction of the x-axis.

Line l_2 is perpendicular to l_1

(a) Find the gradient of l_1

1

(b) Find the equation of l_2 given that it passes through the point $(0, -\sqrt{2})$

2

(c) Determine where l_2 crosses the x-axis.

2

7. A circle touches both the x-axis and the y-axis and has its centre on the line $y = x$

The centre is at a distance of $\sqrt{2}$ units from the origin.

Find the two possible equations for the circle.

5

8. Find the rate of change of $f(x) = \cos\left(3x - \frac{\pi}{6}\right)$ when $x = \frac{\pi}{3}$

4

9. (a) Express $\log_3(9\sqrt{x})$ in the form $a \log_3 x + b$ where a and b are constants.

3

(b) Hence, or otherwise, evaluate $\log_3(9\sqrt{x})$ when $x = \sqrt{3}$

1

10. Two cubic graphs, $y = f(x)$ and $y = g(x)$, where $f(x) = 2x^3 + 3x + 12$ and $g(x) = 2 + 16x^2 - x^3$, are shown in the diagram.

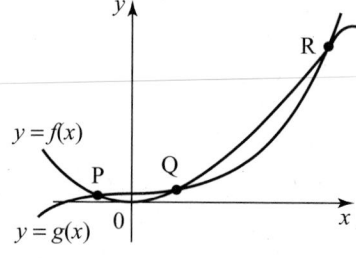

Determine the x-coordinates of each of P, Q and R, the three points of intersection of the two graphs.

8

11. (a) Express $\sqrt{3}\cos x° - \sin x°$ in the form $k\cos(x+a)°$

where $k > 0$ and $0 \le a \le 90$

4

(b) Hence solve the equation $\sqrt{3}\cos x° - \sin x° = 1$ for $0 < x < 360$

3

12. Solve the equation $\sin\theta\,(\sin\theta - 1) = \cos^2\theta$ for $\dfrac{\pi}{2} < \theta < \dfrac{3\pi}{2}$

5

[END OF QUESTION PAPER]

Higher Maths Practice Papers

Practice Papers for SQA Exams

Exam A
Paper 2
Non-calculator

Fill in these boxes and read what is printed below.

Full name of centre

Town

Forename(s)

Surname

Total marks — 70

Attempt ALL questions.

You may use a calculator.

Full credit will be given only to solutions which contain appropriate working.

State the units for your answer where appropriate.

Scotland's leading educational publishers

FORMULAE LIST

Circle:

The equation $x^2 + y^2 + 2gx + 2fy + c = 0$ represents a circle centre $(-g, -f)$ and radius $\sqrt{g^2 + f^2 - c}$

The equation $(x - a)^2 + (y - b)^2 = r^2$ represents a circle centre (a, b) and radius r

Scalar Product: $a.b = |a||b| \cos \theta$, where θ is the angle between a and b

or $a.b = a_1b_1 + a_2b_2 + a_3b_3$ where $a = \begin{pmatrix} a_1 \\ a_2 \\ a_3 \end{pmatrix}$ and $b = \begin{pmatrix} b_1 \\ b_2 \\ b_3 \end{pmatrix}$

Trigonometric formulae:

$$\sin(A \pm B) = \sin A \cos B \pm \cos A \sin B$$

$$\cos(A \pm B) = \cos A \cos B \mp \sin A \sin B$$

$$\sin 2A = 2\sin A \cos A$$

$$\cos 2A = \cos^2 A - \sin^2 A$$

$$= 2\cos^2 A - 1$$

$$= 1 - 2\sin^2 A$$

Table of standard derivatives:

$f(x)$	$f'(x)$
$\sin ax$	$a \cos ax$
$\cos ax$	$-a \sin ax$

Table of standard integrals:

$f(x)$	$\int f(x)\, dx$
$\sin ax$	$-\dfrac{1}{a} \cos ax + C$
$\cos ax$	$\dfrac{1}{a} \sin ax + C$

1. The diagram shows a cubic curve with equation

$$y = x^2 - \frac{1}{3}x^3$$

A tangent PQ to the curve has point of contact $M\,(3, 0)$.

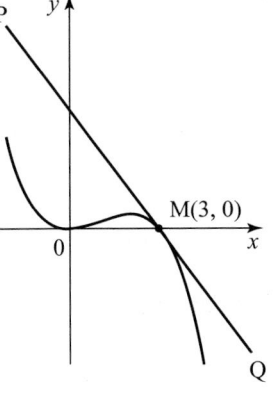

(a) Find the equation of PQ. **4**

A circle has equation $x^2 + y^2 - 4x - 26y + 163 = 0$

(b) Show that PQ is also a tangent to this circle and find the coordinates of the point of contact N. **6**

(c) Find the ratio in which the y-axis cuts the line MN. **3**

2. (a) Find the stationary points on the curve with equation $y = x^3 - 3x^2 + 4$ and justify their nature. **7**

(b) (i) Show that $(x + 1)(x - 2)^2 = x^3 - 3x^2 + 4$

(ii) Hence sketch the graph of $y = x^3 - 3x^2 + 4$ **4**

3. Triangle PQR has coordinates $P(-3, -4)$, $Q(-3, 4)$ and $R(5, 12)$.

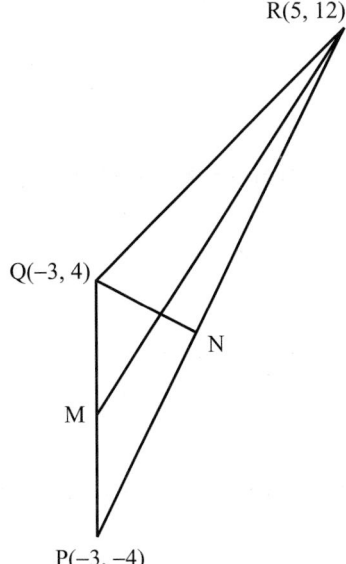

(a) Find the equation of the median MR. **3**

(b) Find the equation of the altitude NQ. **3**

(c) Median MR and altitude NQ intersect at point S. Find the coordinates of S. **3**

(d) The point $T\,(2, 9)$ lies on QR. Show that ST is parallel to PR. **2**

4. 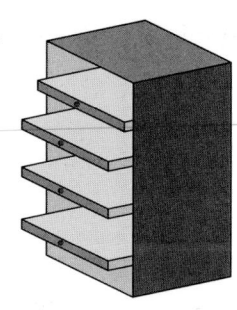 This set of drawers is being 'modelled' on a computer software design package as a cuboid as shown. The edges of the cuboid are parallel to the x, y and z-axes. Three of the vertices are $P(-1, -1, -1)$, $S(-1, 4, -1)$ and $V(3, 4, 5)$.

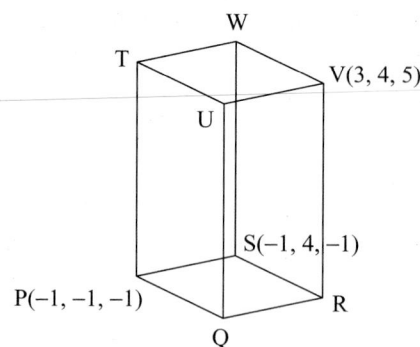

(a) Write down the lengths of PQ, QR and RV.

1

(b) Write down the components of $\overrightarrow{VS}$ and $\overrightarrow{VP}$ and hence calculate the size of angle PVS.

7

5. The islanders living in Tarbert on the island of Harris are planning to build a new sewage processing plant. Central to the plant is the seepage pit which allows most of the week's sewage to seep harmlessly through the soil and drain away. Sewage is pumped into the pit at the start of each week. These are two possible sites with the following specifications:

	Seepage Rate	Pumping capacity
Upland Site:	65% of 1 week's sewage	2000 litres at start of week
Lowland Site:	75% of 1 week's sewage	2500 litres at start of week

(a) Write down a recurrence relation for each site. Use u_n to represent the amount of litres of sewage stored at the Upland site immediately after pumping at the start of the nth week, and let v_n be the equivalent volume at the Lowland site. Clearly label each relation with the site name.

2

(b) The size of the storage tank at each site is determined by the maximum volume of sewage that will remain at the site in the long term. Which site requires the smaller tank in the long term?

4

6. Atmospheric pressure decreases exponentially as you rise above sea-level. It is known that the atmospheric pressure, P(h), at a height h kilometres above sea level is given by P(h) $= P_0e^{-kh}$ where P_0 is the pressure at sea-level ($h = 0$).

(a) Given that at a height of 4·95 km the atmospheric pressure is half that at sea-level, calculate the value of k correct to 4 decimal places.

3

(b) Mount Everest is 8850 metres high. What is the percentage decrease in air pressure at the top of Mount Everest, compared to the pressure at sea-level?

2

7. The diagram shows the curve with equation $y = 6 + 4x - x^2$ and the straight line with equation $y = x + 2$. The line intersects the curve at points S and T as shown

(a) Calculate the exact value of the area enclosed by the curve and the line.

7

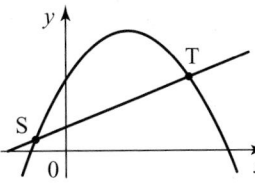

(b) A point $P(x, y)$ lies on the curve between S and T. It is known that the area of triangle PST (shaded in the diagram) is given by:

$$A(x) = -\frac{5}{2}x^2 + \frac{15}{2}x + 10$$

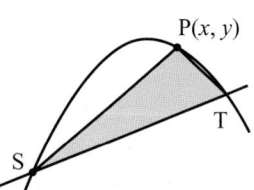

Calculate the maximum value of this area and hence determine what fraction this maximum value is of the area enclosed by the curve and the line from part (a).

4

8. In right-angled triangle PQR, RS is the bisector of angle PRQ.

$PR = 5$ units and $PQ = 12$ units. Show that the exact value of $\cos \theta$ is $\dfrac{3\sqrt{3}}{13}$

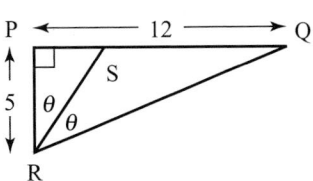

5

[END OF QUESTION PAPER]

Higher Maths Practice Papers

Practice Papers for SQA Exams

Exam B
Paper 1
Non-calculator

Fill in these boxes and read what is printed below.

Full name of centre

Town

Forename(s)

Surname

Total marks — 60

Attempt ALL questions.

You may NOT use a calculator.

Full credit will be given only to solutions which contain appropriate working.

State the units for your answer where appropriate.

Leckie✕Leckie

Scotland's leading educational publishers

FORMULAE LIST

Circle:

The equation $x^2 + y^2 + 2gx + 2fy + c = 0$ represents a circle centre $(-g, -f)$ and radius $\sqrt{g^2 + f^2 - c}$

The equation $(x - a)^2 + (y - b)^2 = r^2$ represents a circle centre (a, b) and radius r

Scalar Product: $\quad a.b = |a||b| \cos \theta$, where θ is the angle between a and b

or $\quad a.b = a_1b_1 + a_2b_2 + a_3b_3$ where $a = \begin{pmatrix} a_1 \\ a_2 \\ a_3 \end{pmatrix}$ and $b = \begin{pmatrix} b_1 \\ b_2 \\ b_3 \end{pmatrix}$

Trigonometric formulae:

$$\sin (A \pm B) = \sin A \cos B \pm \cos A \sin B$$

$$\cos (A \pm B) = \cos A \cos B \mp \sin A \sin B$$

$$\sin 2A = 2\sin A \cos A$$

$$\cos 2A = \cos^2 A - \sin^2 A$$

$$= 2\cos^2 A - 1$$

$$= 1 - 2\sin^2 A$$

Table of standard derivatives:

$f(x)$	$f'(x)$
$\sin ax$	$a \cos ax$
$\cos ax$	$-a \sin ax$

Table of standard integrals:

$f(x)$	$\int f(x)\, dx$
$\sin ax$	$-\dfrac{1}{a} \cos ax + C$
$\cos ax$	$\dfrac{1}{a} \sin ax + C$

1. Find $\int 2x^2 + 3\,dx$

3

2. A function is defined on the set of real numbers by:

$$f(x) = \frac{1}{x+1} \quad (x \neq -1)$$

(a) Find an expression for $f^{-1}(x)$

2

(b) State a suitable domain for f^{-1}

1

3. The diagram shows a sketch of part of the graph $y = \log_3 x$

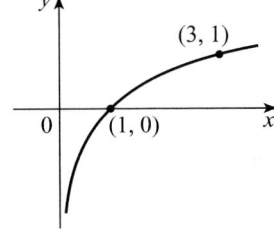

(a) Draw a sketch of the related graph
$$y = 2\log_3(x + 3)$$
showing clearly where it crosses the x-axis and y-axis.

2

(b) Hence, sketch the graph $y = \log_3 \frac{1}{(x+3)^2}$

2

4. The diagram shows a cube with edge length 1 unit. Vectors p and q are represented by the line segments as shown.

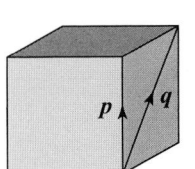

(a) Evaluate: (i) $p.p$ (ii) $q.q$ (iii) $p.q$

3

(b) If r is the resultant of $2p + q$ evaluate $r.r$

3

(c) Hence, write down the value of $|r|$

1

5. (a) Find the exact value of $\cos\left(A + \frac{5\pi}{6}\right) - \sin\left(A + \frac{4\pi}{3}\right)$ by expanding and simplifying.

3

(b) Hence, show that $\cos\left(\frac{11\pi}{12}\right) = \sin\left(\frac{17\pi}{12}\right)$

1

6. If $p = \begin{pmatrix} -1 \\ 2 \\ 3 \end{pmatrix}$ and $q = \begin{pmatrix} 1 \\ 3 \\ 2 \end{pmatrix}$

(a) Write down the components of $p + q$ and $p - q$

1

(b) Hence, show that $p + q$ and $p - q$ are perpendicular.

2

7. Molten metal in a cooling chamber cools according to the law $T_t = T_0 \times 10^{-kt}$ where T_0 is the initial temperature and T_t is the temperature after t minutes. All temperatures are measured in °C.

    (a) A metal at 3000°C takes 20 minutes to cool to 300°C. Calculate the value of k.

                                    2

    (b) What will the temperature of the metal be after a further 20 minutes? (Show clearly your reasoning.)

                                    2

8. (a) Show that $(x + 4)$ is a factor of $x^3 + 6x^2 - 32$

                                    3

    (b) Hence or otherwise, factorise $x^3 - 6x^2 - 32$ fully.

                                    2

9. (a) Find $\int \sqrt{1 + 3x}\, dx$

                                    3

    (b) Hence, show that $\int_0^1 \sqrt{1 + 3x}\, dx = \dfrac{14}{9}$

                                    2

10. The terms of a sequence are generated by the recurrence relation
$$u_{n+1} = pu_n - 1 \text{ with } u_0 = 3$$

    (a) Express u_1 and u_2 in terms of p.

                                    2

    (b) Given that $u_2 = 1$ find the value of p that generates a sequence with a limit.

                                    3

    (c) Calculate the value of this limit.

                                    2

11. The diagram shows a sketch of the curve with equation $y = \dfrac{1}{16}x^4 - \dfrac{1}{8}x^2 + x$. The line $y = x + c$ is a tangent to this curve.

Find the possible values for c. For each value, find the coordinates of the point of contact of the tangent.

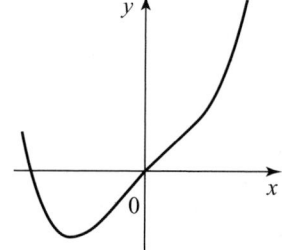

                                    7

12. Functions f and g are defined by $f(x) = 2x - 1$ and $g(x) = \log_{12} x$ on suitable domains.

    (a) Show that the equation $f(g(x)) + g(f(x)) = 0$ has a solution $x = 2$.

                                    6

    (b) Show that the equation has no other real solutions.

                                    2

[END OF QUESTION PAPER]

Higher Maths Practice Papers

Practice Papers for SQA Exams

Exam B
Paper 2
Non-calculator

Fill in these boxes and read what is printed below.

Full name of centre

Town

Forename(s)

Surname

Total marks — 70

Attempt ALL questions.

You may NOT use a calculator.

Full credit will be given only to solutions which contain appropriate working.

State the units for your answer where appropriate.

Scotland's leading educational publishers

FORMULAE LIST

Circle:

The equation $x^2 + y^2 + 2gx + 2fy + c = 0$ represents a circle centre $(-g, -f)$ and radius $\sqrt{g^2 + f^2 - c}$

The equation $(x - a)^2 + (y - b)^2 = r^2$ represents a circle centre (a, b) and radius r

Scalar Product: $a.b = |a||b| \cos \theta$, where θ is the angle between a and b

or $a.b = a_1b_1 + a_2b_2 + a_3b_3$ where $a = \begin{pmatrix} a_1 \\ a_2 \\ a_3 \end{pmatrix}$ and $b = \begin{pmatrix} b_1 \\ b_2 \\ b_3 \end{pmatrix}$

Trigonometric formulae: $\sin (A \pm B) = \sin A \cos B \pm \cos A \sin B$

$$\cos (A \pm B) = \cos A \cos B \mp \sin A \sin B$$

$$\sin 2A = 2\sin A \cos A$$

$$\cos 2A = \cos^2 A - \sin^2 A$$

$$= 2\cos^2 A - 1$$

$$= 1 - 2\sin^2 A$$

Table of standard derivatives:

$f(x)$	$f'(x)$
$\sin ax$	$a \cos ax$
$\cos ax$	$-a \sin ax$

Table of standard integrals:

$f(x)$	$\int f(x)\, dx$
$\sin ax$	$-\dfrac{1}{a} \cos ax + C$
$\cos ax$	$\dfrac{1}{a} \sin ax + C$

1. A function f is defined by the formula $f(x) = x^3 + 3x^2 - 4$

   (a) Find the coordinates of the stationary points on the graph with equation $y = f(x)$ and determine their nature.

6

   (b) Given that $x^3 + 3x^2 - 4 = (x + 2)^2 (x - 1)$ find the coordinates of the points where the curve $y = f(x)$ crosses the x and y-axes and hence sketch the curve.

4

2. Three circles have equations as follows:

      Circle A: $x^2 + y^2 + 4x - 6y + 5 = 0$

      Circle B: $(x - 2)^2 + (y + 1)^2 = 2$

      Circle C: $(x - 2)^2 + (y + 1)^2 = 40$

   (a) (i) State the centre of circle A.

1

      (ii) Show that the radius of circle A is $2\sqrt{2}$.

1

   (b) (i) Calculate the distance between the centres of circles A and B writing your answer as a surd in its simplest form.

2

      (ii) Hence show that circles A and B do not intersect.

2

   (c) Circles A and C intersect at points P and Q. Chord PQ has equation $y = x + 5$. Find the coordinates of points P and Q if P lies to the left of Q.

5

3. The diagram shows kite $ABCD$ with diagonal AC drawn. The vertices of the kite are $A(-5, 2)$, $B(-3, 8)$, $C(3, 6)$ and $D(5, -8)$.

The dotted line shows the perpendicular bisector of AB.

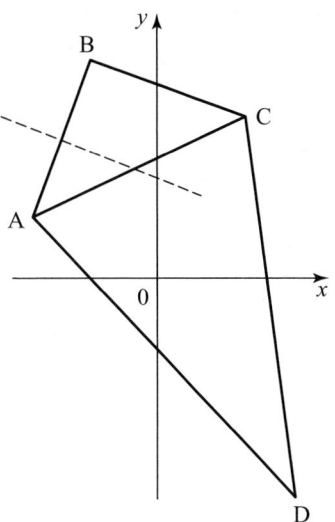

   (a) Show that the perpendicular bisector of AB has equation $3y + x = 11$

4

   (b) Find the equation of the median from C in triangle ACD.

3

   (c) The perpendicular bisector of AB and the median from C in triangle ACD meet at the point S. Find the coordinates of S.

3

4. Solve the equation

$$3 \cos 2x° + 9 \cos x° = \cos^2 x° - 7 \text{ for } 0 \le x < 360$$

5

5. $OABC, DEFG$ is cuboid.

The vertex F is the point (5, 6, 2).

M is the midpoint of DG.

N divides AB in the ratio 1:2.

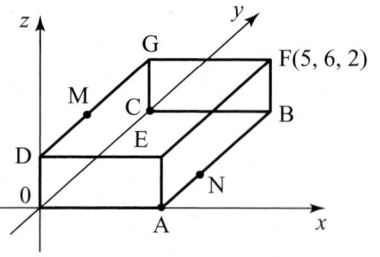

(a) Find the coordinates of M and N.

2

(b) Write down the components of $\overrightarrow{MB}$ and $\overrightarrow{MN}$.

2

(c) Find the size of angle BMN.

5

6. The diagram below shows the graphs $y = f(x)$ and $y = g(x)$ where $f(x) = m \sin x$ and $g(x) = n \cos x$

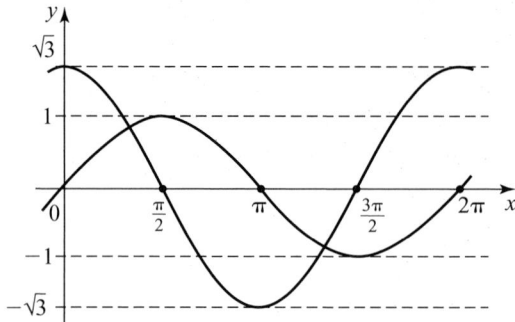

(a) Write down the values of m and n.

2

(b) Write $f(x) - g(x)$ in the form $k \sin(x - a)$ where $k > 0$ and $0 < a < \dfrac{\pi}{2}$

4

(c) Hence find, in the interval $0 \le x \le \pi$ the x-coordinate of the point on the curve $y = f(x) - g(x)$ where the gradient is 2.

2

7. The diagram shows the graph with equation $y = x^4 - 1$

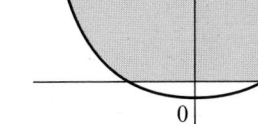

The graph has the y-axis as an axis of symmetry.

The shaded area lies between the curve, the x-axis and the line $y = 15$.

Calculate the exact value of the shaded area.

8

8. The diagram shows a parabola with equation $y = f(x)$ passing through the points $(-3, 0)$, $(0, 9)$ and $(3, 0)$.

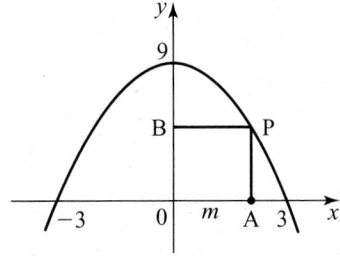

$OAPB$ is a rectangle with A and B lying on the axes and P lying on the parabola as shown. $OA = m$, $0 < m < 3$

(a) If $f(x)$ is of the form $-x^2 + a$ where a is a constant, determine the value of a.

1

(b) Show that $AP = 9 - m^2$

1

(c) Find the value of m for which the area of the rectangle has a maximum.

6

(d) Find the exact value of this maximum area.

1

[END OF QUESTION PAPER]

Practice Paper C

Higher Maths Practice Papers

Practice Papers for SQA Exams

Exam C
Paper 1
Non-calculator

Fill in these boxes and read what is printed below.

Full name of centre

Town

Forename(s)

Surname

Total marks — 60

Attempt ALL questions.

You may NOT use a calculator.

Full credit will be given only to solutions which contain appropriate working.

State the units for your answer where appropriate.

Leckie × Leckie

Scotland's leading educational publishers

FORMULAE LIST

Circle:

The equation $x^2 + y^2 + 2gx + 2fy + c = 0$ represents a circle centre $(-g, -f)$ and radius $\sqrt{g^2 + f^2 - c}$

The equation $(x - a)^2 + (y - b)^2 = r^2$ represents a circle centre (a, b) and radius r

Scalar Product: $a.b = |a||b| \cos \theta$, where θ is the angle between a and b

$$\text{or} \quad a.b = a_1 b_1 + a_2 b_2 + a_3 b_3 \text{ where } a = \begin{pmatrix} a_1 \\ a_2 \\ a_3 \end{pmatrix} \text{ and } b = \begin{pmatrix} b_1 \\ b_2 \\ b_3 \end{pmatrix}$$

Trigonometric formulae:

$$\sin (A \pm B) = \sin A \cos B \pm \cos A \sin B$$

$$\cos (A \pm B) = \cos A \cos B \mp \sin A \sin B$$

$$\sin 2A = 2\sin A \cos A$$

$$\cos 2A = \cos^2 A - \sin^2 A$$

$$= 2\cos^2 A - 1$$

$$= 1 - 2\sin^2 A$$

Table of standard derivatives:

$f(x)$	$f'(x)$
$\sin ax$	$a \cos ax$
$\cos ax$	$-a \sin ax$

Table of standard integrals:

$f(x)$	$\int f(x)\,dx$
$\sin ax$	$-\dfrac{1}{a} \cos ax + C$
$\cos ax$	$\dfrac{1}{a} \sin ax + C$

1. The diagram shows a crystal of Alum.
Three vertices of the crystal have coordinates
$A(5, 16, 25)$, $B(7, 14, 21)$ and $C(10, 11, 15)$.

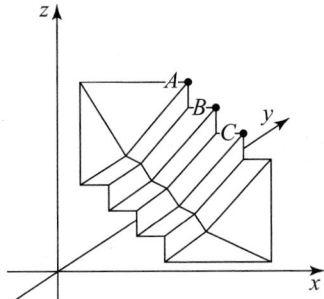

(a) Show that the points A, B and C are collinear.

4

(b) Find the ratio in which B divides AC.

1

2. Three functions f, g and h are defined on the set of real numbers by
$f(x) = 4x - 1$, $g(x) = \frac{1}{2}x + 1$ and $h(x) = \frac{1}{2}(x - 3)$

(a) Find a formula for $p(x) = f(g(x))$

2

(b) Find a formula for $h(p(x))$

2

(c) Describe the relationship between the functions h and p.

1

3. The diagram shows part of the cubic graph $y = f(x)$

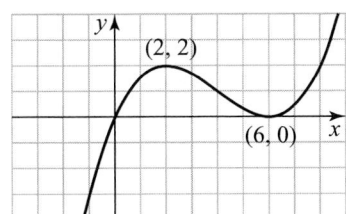

Make a sketch of the graph $y = -f(\frac{1}{2}x)$ showing clearly any turning points and
where the graph crosses the x-axis.

3

4. Find the rate of change of v with respect to t where $v = \sqrt{\sin t}$

3

5. Two facts are known about the polynomial $f(x) = x^4 - 5x^3 + ax^2 + bx - 2$:
• $(x - 2)$ is a factor
• when it is divided by $x + 1$ the remainder is 12.

Use these two facts to calculate the values of a and b.

5

6. Find the equation of the line through the point $(-3, 2)$ which is perpendicular to the
line with equation $2x - 3y + 1 = 0$

3

7. The recurrence relation $u_{n+1} = 0.3u_n + k$ generates a sequence of terms with $u_0 = 10$

(a) If $u_2 = 7.4$ calculate the value of the constant k.

3

(b) Give a reason why the sequence has a limit L.

1

(c) Calculate the exact value of L.

2

8. The point $(\frac{\pi}{6}, -1)$ lies on the graph $y = f(x)$.
If $f'(x) = \sin 2x$ find $f(x)$.

3

9. (a) Find the stationary points on the curve with equation $y = x^3 - 3x^2 - 24x - 28$ and justify their nature.

7

(b) The curve intersects the x-axis at $(7, 0)$. Sketch the curve.

2

10. $ABCD$ is a rectangle with point P lying on side AB, 1 unit from A. $AD = 2$ units

The dotted line PD on the diagram shows the bisector of angle APC with angle $APD =$ angle $DPC = y°$

1

(a) Find $\theta°$ in terms of $y°$

(b) Hence find the exact value of $\sin \theta°$

6

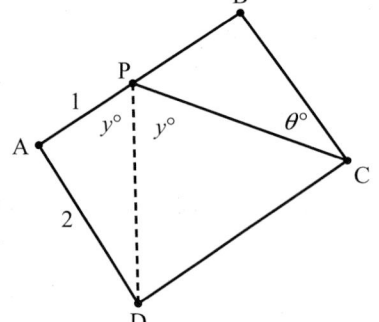

11. Solve the equation $\log_{\sqrt{2}} x - \log_{\sqrt{2}} 2 = 2$

4

12. The diagram shows part of the graph $y = a \sin (x + b) + c$
It has a turning point at $(\frac{\pi}{3}, -1)$ and a maximum value of 2.

(a) Calculate the values of a, b and c.

3

(b) Find the coordinates of the point A where the graph crosses the y-axis.

2

(c) The line $y = \frac{5}{4}$ intersects the graph at point B as shown in the diagram. Find the exact value of the x-coordinate of point B.

2

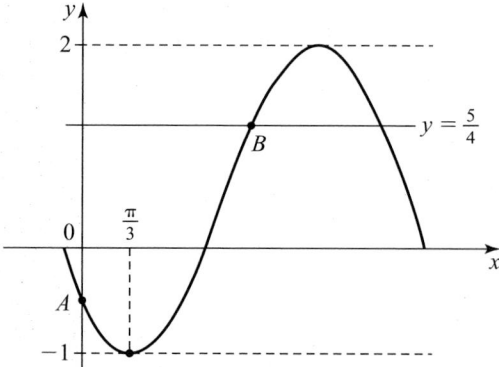

[END OF QUESTION PAPER]

Higher Maths Practice Papers

Practice Papers for SQA Exams

Exam C
Paper 2
Non-calculator

Fill in these boxes and read what is printed below.

Full name of centre

Town

Forename(s)

Surname

Total marks — 70

Attempt ALL questions.

You may NOT use a calculator.

Full credit will be given only to solutions which contain appropriate working.

State the units for your answer where appropriate.

Scotland's leading educational publishers

FORMULAE LIST

Circle:

The equation $x^2 + y^2 + 2gx + 2fy + c = 0$ represents a circle centre $(-g, -f)$ and radius $\sqrt{g^2 + f^2 - c}$

The equation $(x - a)^2 + (y - b)^2 = r^2$ represents a circle centre (a, b) and radius r

Scalar Product: $\quad a.b = |a||b| \cos \theta$, where θ is the angle between a and b

$$\text{or} \quad a.b = a_1 b_1 + a_2 b_2 + a_3 b_3 \text{ where } a = \begin{pmatrix} a_1 \\ a_2 \\ a_3 \end{pmatrix} \text{ and } b = \begin{pmatrix} b_1 \\ b_2 \\ b_3 \end{pmatrix}$$

Trigonometric formulae:

$$\sin(A \pm B) = \sin A \cos B \pm \cos A \sin B$$

$$\cos(A \pm B) = \cos A \cos B \mp \sin A \sin B$$

$$\sin 2A = 2\sin A \cos A$$

$$\cos 2A = \cos^2 A - \sin^2 A$$

$$= 2\cos^2 A - 1$$

$$= 1 - 2\sin^2 A$$

Table of standard derivatives:

$f(x)$	$f'(x)$
$\sin ax$	$a \cos ax$
$\cos ax$	$-a \sin ax$

Table of standard integrals:

$f(x)$	$\int f(x)\, dx$
$\sin ax$	$-\dfrac{1}{a} \cos ax + C$
$\cos ax$	$\dfrac{1}{a} \sin ax + C$

1. Solve the equation $\sin 2x - \sqrt{3} \sin x = 0$ for $0 \leq x \leq 2\pi$

5

2. (a) Express $3 \sin x° - \cos x°$ in the form $k \sin (x - a)°$ where $k > 0$ and $0 \leq a \leq 90$

4

   (b) Hence, solve the equation $3 \sin x° - \cos x° = 1$ for $0 \leq x \leq 90$

3

3. The vectors $\overrightarrow{BA}$ and $\overrightarrow{BC}$ have components $\begin{pmatrix} -2 \\ 3 \\ 5 \end{pmatrix}$ and $\begin{pmatrix} 1 \\ -1 \\ 3 \end{pmatrix}$ respectively.

   Calculate the size of angle ABC.

5

4. Prove that for all values of c the equation $x^2 - 2x + c^2 + 2 = 0$ has no real roots.

4

5. The diagram shows the graph with equation $y = \frac{1}{3}x^3 - 2x^2$

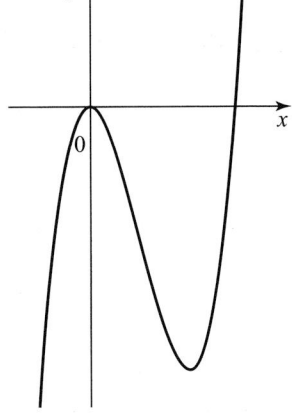

   (a) A tangent to this curve has gradient -4. Find the x-coordinate of the point of contact.

5

   (b) Hence, find the equation of this tangent.

2

6. A drug is given to a patient. The concentration, C_t milligrams per millilitre (mg/ml), of the drug in the patient's blood t hours after it is administered is given by the formula:

$$C_t = C_o e^{-\frac{t}{4}}$$

   where C_0 is the concentration in the blood immediately after the drug was administered.

   (a) If the concentration is 3·5 mg/ml after 3 hours, what was the concentration of the drug just after it was administered?

3

   (b) In general, after a dose of this drug has been administered, how long does it take for the initial concentration to halve?

4

7. The line L is a tangent to the circle with centre C_1 and equation:

$$x^2 + y^2 - 4x - 6y + 8 = 0$$

The point of contact A has coordinates (1, 5).

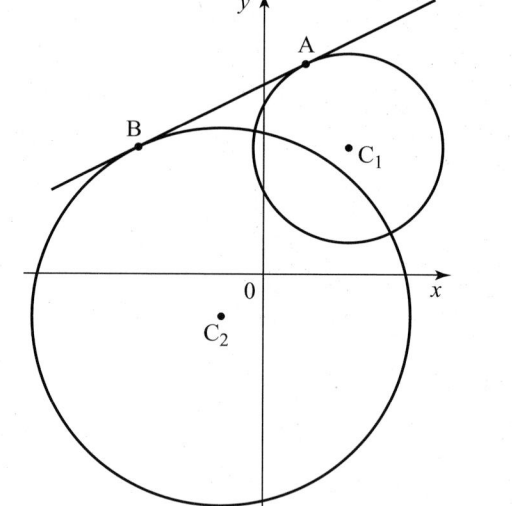

(a) Show that the equation of line L is $2y - x = 9$

4

The circle with centre C_2 has equation:

$$x^2 + y^2 + 2x + 2y - 18 = 0$$

(b) Show that line L is also a tangent to this circle.

5

(c) If B is the point of contact, find the exact length of AB.

2

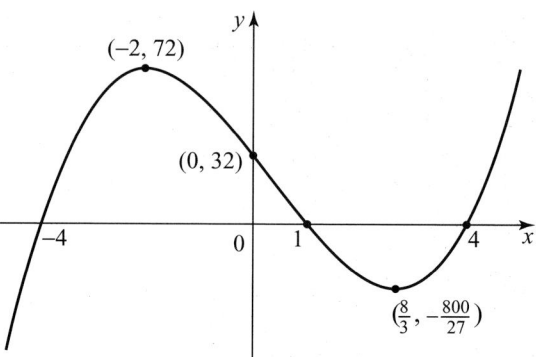

8. The graph shows a cubic function with equation $y = f(x)$

The graph has stationary points at:

$(-2, 72)$ and $\left(\dfrac{8}{3}, -\dfrac{800}{27} \right)$

The graph intersects the axes at the points $(-4, 0)$, $(1, 0)$, $(4, 0)$ and $(0, 32)$.

Sketch the graph of $y = f'(x)$

3

9. An open box is in the shape of a cuboid and was made from a sheet of tin.

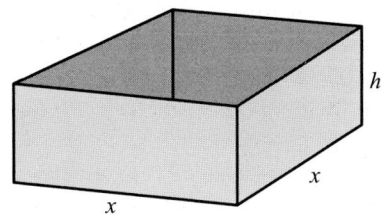

The box has a square base of side x cm and a height of h cm.

The volume of the box is $62\frac{1}{2}$ cm³

(a) Show that the area, A cm², of tin required to make the box is given by:

$A(x) = \frac{250}{x} + x^2$

3

(b) Find the value of x for which this area is a minimum.

5

10. The diagram shows a rectangular metal plate with dimensions 5 cm × 6 cm.

The plate has a parabolic section removed from it.

The equation of the parabola used to make this section is $y = x^2 - 6x + 10$

The scale of the diagram is 1 unit = 1 cm

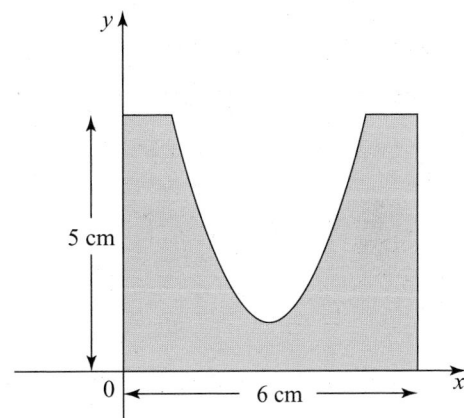

Find the shaded area, in square centimetres, of the metal plate.

8

11. The diagram shows the graph $y = g(x)$ where $g(x) = x^2(x - 1)$.

A function f is such that $y = f'(x)$ is identical to the graph $y = g(x)$ as shown in the diagram.

If $f(2) = \frac{1}{3}$, find the formula for $f(x)$.

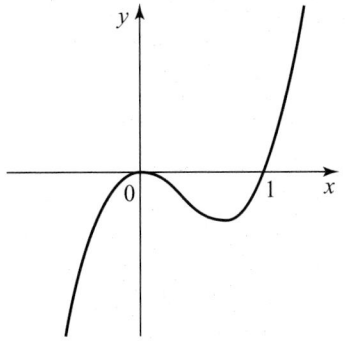

5

[END OF QUESTION PAPER]

Answers

Exam A: Paper 1

Q1. (a)
$\cos(x - 30)° = \cos x° \cos 30° + \sin x° \sin 30°$ ✔

1 mark

Q1. (b)
Let $x = 45$ giving ✔
$\cos 15° = \cos(45 - 30)°$
$\quad = \cos 45° \cos 30° + \sin 45° \sin 30°$ ✔
$\quad = \dfrac{1}{\sqrt{2}} \times \dfrac{\sqrt{3}}{2} + \dfrac{1}{\sqrt{2}} \times \dfrac{1}{2}$
$\quad = \dfrac{\sqrt{3}}{2\sqrt{2}} + \dfrac{1}{2\sqrt{2}} = \dfrac{\sqrt{3}+1}{2\sqrt{2}}$ ✔

3 marks

Expansion
- You are given this formula during your exam:
$\cos(A \pm B) = \cos A \cos B \mp \sin A \sin B$

Use of $x = 45$
- 1st mark is for evidence of $45 - 30$

Exact values
- $\sin 45° = \cos 45° = \dfrac{1}{\sqrt{2}}$

  $\sin 30° = \dfrac{1}{2} \quad \cos 30° = \dfrac{\sqrt{3}}{2}$
- You should be able to work out these exact values by drawing the appropriate triangles.

Simplification
- A single fraction is required for this mark.
- Addition and multiplication of fractions:

  $\frac{a}{b} \times \frac{c}{d} = \frac{ac}{bd} \qquad \frac{a}{c} + \frac{b}{c} = \frac{a+b}{c}$

HMSG: p 24, p 30

Q2.
Compare $(2k + 3)x^2 - kx - 1 = 0$
with: $\qquad ax^2 + bx + c = 0$
This gives: $a = 2k + 3$, $b = -k$ and $c = -1$
Discriminant $= b^2 - 4ac$
$\qquad = (-k)^2 - 4(2k + 3) \times (-1)$ ✔
$\qquad = k^2 + 4(2k + 3)$
$\qquad = k^2 + 8k + 12$ ✔
For real roots set Discriminant ≥ 0 ✔
$\Rightarrow k^2 + 8k + 12 \geq 0 \Rightarrow (k + 6)(k + 2) \geq 0$
$\Rightarrow k \leq -6$ or $k \geq -2 \qquad$ Graph:

So the roots are real
for $k \leq -6$ or $k \geq -2$

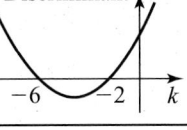

✔

4 marks

Substitution
- Correct identification of a, b and c and substitution into $b^2 - 4ac$
- Be careful to include the negative signs

Simplification
- Squaring a negative gives a positive answer
- Note that you have $-4 \times (-1) = 4$

Condition
- You must clearly state the condition for real roots: Discriminant ≥ 0

Solution
- Notice $(k + 6)(k + 2) = 0 \Rightarrow k = -6$ or $k = -2$ and for $(k + 6)(k + 2) \geq 0$ you are looking for where the graph (showing the values of the discriminant) lies above or on the k-axis.

HMSG: pp 46–48

Q3.

$\displaystyle\int \dfrac{2x^5 - 3}{3x^4}\, dx = \int \dfrac{2x^5}{3x^4} - \dfrac{3}{3x^4}\, dx$

$\displaystyle = \int \dfrac{2x}{3} - \dfrac{1}{x^4}\, dx = \int \dfrac{2x}{3} - x^{-4}\, dx$ ✔ ✔

$\displaystyle = \dfrac{2x^2}{3 \times 2} - \dfrac{x^{-3}}{-3} + C = \dfrac{x^2}{3} + \dfrac{x^{-3}}{3} + C$ ✔

$\displaystyle = \dfrac{1}{3}x^2 + \dfrac{1}{3x^3} + C$ ✔

4 marks

Preparation
- This mark is for correctly preparing for differentiation: terms $\frac{2}{3}x$ and x^{-4} appear after splitting the fraction into two terms.

Integration of 1st term
- Correct integration of $\frac{2}{3}x$

Integration of 2nd term
- Correct integration of x^{-4}
- Increasing -4 by 1 gives -3 not -5

Simplification & constant
- Forget the constant C and you lose the mark!

HMSG: pp 68–69

Q4. (a)

$h(x) = g(f(x))$

$\quad = g(\frac{1}{2}x + 1)$ ✓

$\quad = 3\left(\frac{1}{2}x + 1\right) - \frac{1}{2}$

$\quad = \frac{3}{2}x + 3 - \frac{1}{2} = \frac{3}{2}x + \frac{5}{2}$ ✓

Substitution
- Replacement of $f(x)$ by $\frac{1}{2}x + 1$ for 1st mark.

Substitution & simplification
- Notice that g acts on $\frac{1}{2}x + 1$ by multiplying it by 3 then subtracting $\frac{1}{2}$
- The expression must be simplified to gain this mark.

2 marks

Q4. (b)

$h(x) = \frac{3}{2}x + \frac{5}{2}$

Suppose $b = \frac{3}{2}a + \frac{5}{2}$

$\Rightarrow 2b = 3a + 5 \Rightarrow 2b - 5 = 3a$ ✓

$\Rightarrow \frac{2b-5}{3} = a \Rightarrow a = \frac{2}{3}b - \frac{5}{3}$

So $h^{-1}(x) = \frac{2}{3}x - \frac{5}{3}$ ✓

$\Rightarrow p = \frac{2}{3}$ and $q = -\frac{5}{3}$ ✓

Change the subject
- Evidence that you have used a correct strategy is required to gain this mark.

Inverse
- When $x = a$ goes in to $h(x)$ then b comes out. When b goes in to $h^{-1}(x)$ then a comes out. Start with b in terms of a finish with a in terms of b. Replace b by x for the inverse formula.

Values
- The question asks for the values of p and q.

HMSG: p 10

3 marks

Q5. $\overrightarrow{AB} = \begin{pmatrix} 4 \\ 0 \\ -2 \end{pmatrix}$ with $A\,(-1, 2, 3)$

so $B\,(-1 + 4, 2 + 0, 3 - 2) = B\,(3, 2, 1)$ ✓

$\overrightarrow{BC} = \begin{pmatrix} 4 \\ 0 \\ -2 \end{pmatrix} + \begin{pmatrix} 2 \\ 2 \\ 2 \end{pmatrix} = \begin{pmatrix} 6 \\ 2 \\ 0 \end{pmatrix}$

so $C\,(3 + 6, 2 + 2, 1 + 0) = C\,(9, 4, 1)$ ✓

$\overrightarrow{AB} = -2\overrightarrow{CD} \Rightarrow \overrightarrow{CD} = -\frac{1}{2}\overrightarrow{AB} = -\frac{1}{2}\begin{pmatrix} 4 \\ 0 \\ -2 \end{pmatrix} = \begin{pmatrix} -2 \\ 0 \\ 1 \end{pmatrix}$

so $D\,(9 - 2, 4 + 0, 1 + 1) = D\,(7, 4, 2)$ ✓

$\overrightarrow{DE} = \frac{1}{2}\overrightarrow{BC} = \frac{1}{2}\begin{pmatrix} 6 \\ 2 \\ 0 \end{pmatrix} = \begin{pmatrix} 3 \\ 1 \\ 0 \end{pmatrix}$

so $E\,(7 + 3, 4 + 1, 2 + 0) = E\,(10, 5, 2)$ ✓

1ˢᵗ point
- This solution uses the following result:

  if $P\,(x, y, z)$ and $\overrightarrow{PQ} = \begin{pmatrix} a \\ b \\ c \end{pmatrix}$ then you add the components of $\overrightarrow{PQ}$ to the corresponding coordinates of P to get $Q\,(x + a, y + b, z + c)$

2ⁿᵈ point
- It is usually easier to deal with $\begin{pmatrix} a \\ b \\ c \end{pmatrix}$ rather than $a\mathbf{i} + b\mathbf{j} + c\mathbf{k}$ as is shown in this solution.

3ʳᵈ point
- Using the result: $k\begin{pmatrix} a \\ b \\ c \end{pmatrix} = \begin{pmatrix} ka \\ kb \\ kc \end{pmatrix}$

Answer
- The coordinates of point E are required.

HMSG: pp 34–36, p 39

4 marks

Q6. (a) ✓

$m_{l1} = \tan\frac{\pi}{3} = \sqrt{3}$

Gradient
- Using the result $m = \tan\theta$

1 mark

Q6. (b)

$m_{l1} = \sqrt{3} \Rightarrow m_\perp = -\frac{1}{\sqrt{3}} \Rightarrow m_{l2} = -\frac{1}{\sqrt{3}}$ ✓

Gradient of $l_2 = -\frac{1}{\sqrt{3}}$ and point on l_2 is $(0, -\sqrt{2})$

$\Rightarrow$ equation of l_2 is $y - (-\sqrt{2}) = -\frac{1}{\sqrt{3}}(x - 0)$

$\Rightarrow y + \sqrt{2} = -\frac{1}{\sqrt{3}}x \Rightarrow \sqrt{3}y + \sqrt{6} = -x$

$\Rightarrow \sqrt{3}y + x = -\sqrt{6}$ ✓

Perpendicular gradient
- Using the result $m = \frac{a}{b} \Rightarrow m_\perp = -\frac{b}{a}$

Equation
- If the gradient $= m$ and point on line is (a, b) then the equation is $y - b = m(x - a)$

2 marks

Q6. (c)

For the x-axis intercept set $y = 0$ ✓

so $\sqrt{3} \times 0 + x = -\sqrt{6} \Rightarrow x = -\sqrt{6}$

l_2 crosses the x-axis at the point

$(-\sqrt{6}, 0)$ ✓

2 marks

Strategy
• Evidence of $y = 0$ for this mark

Intercept
• Calculation of $-\sqrt{6}$ will gain this mark

HMSG: pp 78–79

Q7. This diagram shows the two possible positions for the circles:

$x^2 + x^2 = (\sqrt{2})^2 \Rightarrow 2x^2 = 2$ ✓ ✓

$\Rightarrow x^2 = 1 \Rightarrow x = 1$

(positive)

so $C(1, 1)$ and radius $= 1$ ✓

Equation is: $(x - 1)^2 + (y - 1)^2 = 1^2 = 1$ ✓

By symmetry the other circle has centre $(-1, -1)$ and radius $= 1$

Equation is: $(x + 1)^2 + (y + 1)^2 = 1$ ✓

5 marks

Strategy
• Evidence of use of Pythagoras' Theorem to calculate the radius and centre.
• It is important in questions like this one to draw a diagram.

Calculation
• Correct calculation of $x = 1$
• x measures a length so is a positive quantity.

Centre & radius
• A clear statement showing the coordinates of the centre and the value of the radius.

Equation
• The equation of a circle with centre (a, b) and radius $= r$ is $(x - a)^2 + (y - b)^2 = r^2$ In this case $a = 1$, $b = 1$ and $r = 1$
• You must not leave 1^2 but indicate that $1^2 = 1$

2nd equation
• In this case $a = -1$, $b = -1$ and $r = 1$ are the values used in $(x - a)^2 + (y - b)^2 = r^2$

HMSG: p 83

Q8. $f'(x) = -\sin(3x - \frac{\pi}{6}) \times 3$ ✓

$= -3\sin(3x - \frac{\pi}{6})$ ✓

so $f'(\frac{\pi}{3}) = -3\sin(3 \times \frac{\pi}{3} - \frac{\pi}{6})$ ✓

$= -3\sin(\pi - \frac{\pi}{6}) = -3\sin\frac{5\pi}{6}$

$= -3 \times \frac{1}{2} = -\frac{3}{2}$ ✓

4 marks

Strategy
• You should know that 'rate of change' means you should differentiate i.e. find $f'(x)$

Differentiate cos( )
• On your formula sheet you are told: $f(x): \cos ax \qquad f'(x): -a\sin ax$

Chain rule
• You are using the "chain rule" with the factor 3 appearing when you differentiate $3x - \frac{\pi}{6}$

Evaluation
• $\frac{5\pi}{6}$ is a 2nd quadrant angle where $\sin\frac{5\pi}{6}$ is positive and the same as $\sin\frac{\pi}{6} = \frac{1}{2}$

HMSG: pp 64–65

Q9. (a)

$\log_3(9\sqrt{x}) = \log_3(9 \times x^{\frac{1}{2}})$

$= \log_3 9 + \log_3 x^{\frac{1}{2}}$ ✓

$= 2 + \frac{1}{2}\log_3 x$ ✓

$= \frac{1}{2}\log_3 x + 2$ ✓

3 marks

Addition rule
• Correct use of the rule: $\log(mn) = \log m + \log n$

Simplification
• $\log_3 9 = 2$ is equivalent to $3^2 = 9$
• It is useful to read '$\log_3 9$' as 'what power of 3 gives 9?' with the answer: 2

Power rule
• Use of the log rule: $\log a^n = n \log a$ to change $\log_3 x^{\frac{1}{2}}$ to $\frac{1}{2}\log_3 x$

Q9. (b)

$x = \sqrt{3}$ gives $\frac{1}{2}\log_3\sqrt{3} + 2$

$= \frac{1}{2} \times \frac{1}{2} + 2 = \frac{1}{4} + 2 = \frac{9}{4}$ ✓

1 mark

Evaluation
• Read $\log_3\sqrt{3}$ as 'what power of 3 gives $\sqrt{3}$?' the answer being $\frac{1}{2}$

HMSG: p 20

Q10.

For points of intersection:

Solve $f(x) = g(x)$

$\Rightarrow 2x^3 + 3x + 12 = 2 + 16x^2 - x^3$ ✓

$\Rightarrow 3x^3 - 16x^2 + 3x + 10 = 0$ ✓

$$
\begin{array}{c|cccc}
1 & 3 & -16 & 3 & 10 \\
  &   & 3 & -13 & -10 \\
\hline
  & 3 & -13 & -10 & 0
\end{array}
$$ ✓

0 remainder $\Rightarrow$ 1 is a root

So $x - 1$ is a factor ✓

The equation becomes:

$(x - 1)(3x^2 - 13x - 10) = 0$ ✓

$\Rightarrow (x - 1)(3x + 2)(x - 5) = 0$ ✓

$\Rightarrow x - 1 = 0$ or $3x + 2 = 0$ ✓

or $x - 5 = 0$

$x = 1$ or $x = -\dfrac{2}{3}$ or $x = 5$

So the x-coordinates are:

For point P: $-\dfrac{2}{3}$

For point Q: 1

For point R: 5 ✓

Strategy
- Setting the two formulae equal to each other and solving is the strategy in this question.

Cubic equation
- Recognising a cubic and rearranging terms in order gains a mark.

Strategy
- Trying particular values of x. In this case $x = 1$ was successful.

Interpretation
- $x = 1$ is a solution of the equation so $x - 1$ is a factor

Factorising
- Three marks are allocated to this:
1 mark: starting to factorise
$(x - 1)(3x^2 \ldots)$
1 mark: quadratic factor $3x^2 - 13x - 10$
1 mark: completing the factorisation
$(x - 1)(3x + 2)(x - 5)$

Interpretation
- From the diagram you can assign the three values to the points P, Q and R.

HMSG: pp 50–52

8 marks

Q.11. (a)

$\sqrt{3} \cos x° - \sin x° = k \cos(x + a)°$ ✓

$\sqrt{3} \cos x° - \sin x° = k \cos x° \cos a° - k \sin x° \sin a°$

Comparing coefficients of $\cos x°$ and $\sin x°$:

$\left. \begin{array}{l} k \cos a° = \sqrt{3} \\ k \sin a° = 1 \end{array} \right\}$ ✓

Dividing gives:

$\dfrac{k \sin a°}{k \cos a°} = \dfrac{1}{\sqrt{3}} \Rightarrow \tan a° = \dfrac{1}{\sqrt{3}} \Rightarrow a = 30$ ✓

Squaring and adding gives:

$(k \sin a°)^2 + (k \cos a°)^2 = 1^2 + (\sqrt{3})^2$

$\qquad\qquad\qquad = 1 + 3 = 4$

$\Rightarrow k^2(\sin^2 a° + \cos^2 a°) = 4$

$\Rightarrow k^2 \times 1 = 4 \Rightarrow k = 2 \ (k > 0)$ ✓

so $\sqrt{3} \cos x° - \sin x° = 2 \cos(x + 30)°$

Addition formula
- expanding $\cos(x + a)°$ is the essential first step. The formula:
$\cos(A \pm B) = \cos A \cos B \mp \sin A \sin B$
is on your formulae sheet in the exam.

Coefficients

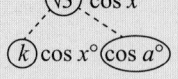 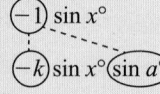

- This gives $k \cos a° = \sqrt{3}$ and $k \sin a° = 1$

Calculation
- Here you are using $\dfrac{\sin a°}{\cos a°} = \tan a°$ along with cancelling the factor k to find angle $a°$.

Calculation
- The result $\sin^2 a° + \cos^2 a° = 1$ leads to the disappearance of $a°$ allowing k to be found.
- k is always positive in this context.

4 marks

Q11. (b)

$\sqrt{3}\cos x° - \sin x° = 1 \Rightarrow 2\cos(x+30)° = 1$

$\Rightarrow \cos(x+30)° = \frac{1}{2}$ ✓

$(x+30)°$ is a 1st or 4th quadrant angle since $\cos(x+30)°$ is positive.

1st quadrant: $\cos^{-1}\frac{1}{2} = 60°$

$\Rightarrow x + 30 = 60 \Rightarrow x = 30$ ✓

4th quadrant:

$x + 30 = 360 - 60 = 300$

$\Rightarrow x = 300 - 30 = 270$ ✓

So for $0 < x < 360$ the solutions are

$x = 30, 270$

3 marks

Set up equation
- Rewrite the equation using the result from (**a**)
- The aim is to rearrange the equation to get:
  cos(an angle) = a number

1st solution
- The 1st quadrant angle is 60° but in this case it's not $x°$ but $(x+30)°$ that equals 60°

2nd solution
- Cosine is positive in the 1st and 4th quadrants. Here is the diagram:

S	A ✓
T	C ✓

HMSG: pp 32–33, p 56

Q12.

$\sin\theta(\sin\theta - 1) = \cos^2\theta$

$\Rightarrow \sin^2\theta - \sin\theta = \cos^2\theta$

$\Rightarrow \sin^2\theta - \sin\theta = 1 - \sin^2\theta$ ✓

$\Rightarrow 2\sin^2\theta - \sin\theta - 1 = 0$ ✓

$\Rightarrow (2\sin\theta + 1)(\sin\theta - 1) = 0$ ✓

$\Rightarrow 2\sin\theta + 1 = 0$ or $\sin\theta = 1$

$\Rightarrow \sin\theta = -\frac{1}{2}$ or $\sin\theta = 1$ ✓

For $\sin\theta = -\frac{1}{2}$

θ is in 3rd or 4th quadrants ($-$ve)

1st quadrant angle is $\frac{\pi}{6}$

However, since $\frac{\pi}{2} < \theta < \frac{3\pi}{2}$

the 4th quadrant solution is not valid.

So $\theta = \pi + \frac{\pi}{6} = \frac{6\pi}{6} + \frac{\pi}{6} = \frac{7\pi}{6}$

For $\sin\theta = 1$

Using the graph gives $\theta = \frac{\pi}{2}$

but since

$\frac{\pi}{2} < \theta < \frac{3\pi}{2}$

this solution is invalid.

So $\theta = \frac{7\pi}{6}$ is the only solution. ✓

5 marks

Strategy
- Looking at the 'make-up' of the equation will lead you to produce an equation in $\sin\theta$ — so use $\sin^2\theta + \cos^2\theta = 1$ to replace $\cos^2\theta$ by $1 - \sin^2\theta$.

'Standard form'
- The equation is a 'quadratic' in $\sin\theta$ and should be arranged in the standard quadratic form: $ax^2 + bx + c = 0$ in this case $a\sin^2\theta + b\sin\theta + c = 0$.

Factorisation
- $2x^2 - x - 1 = (2x+1)(x-1)$ so likewise: $2\sin^2\theta - \sin\theta - 1 = (2\sin\theta + 1)(\sin\theta - 1)$

Solve for $\sin\theta$
- In this case the 'roots' of the quadratic equation are $-\frac{1}{2}$ and 1. These are the possible values for $\sin\theta$.

Solve for θ
- In general for values of $\sin\theta$ or $\cos\theta$ of -1, 0 or 1 you should use the sine or cosine graph to determine the angles. In this case the value $(\frac{\pi}{2})$ is not in the allowed interval $(\frac{\pi}{2} < \theta < \frac{3\pi}{2})$.
- For $\sin\theta = -\frac{1}{2}$ the 4th quadrant solution is not in the allowed interval $(\frac{\pi}{2} < \theta < \frac{3\pi}{2})$.

HMSG: pp 54–55

Exam A: Paper 2

Q1. (a)

$$y = x^2 - \frac{1}{3}x^3 \Rightarrow \frac{dy}{dx} = 2x - x^2 \qquad \checkmark\checkmark$$

At M (3, 0) $x = 3$

So $\dfrac{dy}{dx} = 2 \times 3 - 3^2 = -3$ $\qquad \checkmark$

gradient of tangent is –3.

Point on tangent is (3, 0).

Equation of tangent is:

$$y - 0 = -3(x - 3)$$

$$\Rightarrow y = -3x + 9. \qquad \checkmark$$

4 marks

Strategy
- Knowing to differentiate gains you the 1st mark.

Differentiate
- The correct result gains the 2nd mark.

Calculation
- Where are you on the curve? At the place where $x = 3$, so use $x = 3$ and the gradient formula.

Equation
- Using $y - b = m(x - a)$. In this case $m = -3$ and (a, b) is (3, 0).

HMSG: p 60

Q1. (b)

To find the points of intersection of line and circle

Solve: $\left.\begin{array}{l} y = -3x + 9 \\ x^2 + y^2 - 4x - 26y + 163 = 0 \end{array}\right\}$ $\qquad \checkmark$

Substitute $y = -3x + 9$ in the circle equation: $\qquad \checkmark$

$x^2 + (-3x + 9)^2 - 4x - 26$
$(-3x + 9) + 163 = 0$

$\Rightarrow x^2 + 9x^2 - 54x + 81 - 4x + 78x$
$\qquad -234 + 163 = 0$

$\Rightarrow 10x^2 + 20x + 10 = 0 \qquad \checkmark$

$\Rightarrow 10(x^2 + 2x + 1) = 0$

$\Rightarrow 10(x + 1)(x + 1) = 0$

$\Rightarrow x = -1 \qquad \checkmark$

Since there is only one solution the line is a tangent to the circle. $\qquad \checkmark$

when $x = -1$ $y = -3 \times (-1) + 9 = 12$

So N(-1, 12) is the point of contact. $\qquad \checkmark$

6 marks

Rearrangement
- The form $y = -3x + 9$ is necessary for the subsequent substitution. Do not use $x = \frac{1}{3}y + 3$ as this involves fraction work leading to errors.

Strategy
- Evidence of substitution gains you the strategy mark.

'Standard form'
- This is a quadratic equation and should be written in the 'standard' way, i.e. $ax^2 + bx + c = 0$.

Solution
- Notice that removing a common factor reduces the magnitude of the coefficients and makes the rest of the factorisation easier.

Proof
- A clear statement is required — there was one point of intersection so you have a tangent.

Calculation
- A point was asked for, so you must give the coordinates, $x = -1$ and $y = 12$ is not enough.

HMSG: p 84

Q1. (c)

when $x = 0$ $y = -3 \times 0 + 9 = 9$ ✓

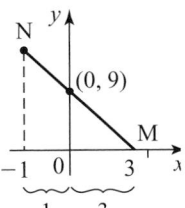

The required
ratio is 3:1 ✓ ✓

3 marks

Strategy
- Knowing where the y-intercept is gains you this strategy mark.

Strategy
- An alternative approach uses vectors.
  If y-intercept is P:
  $\overrightarrow{MP} = \begin{pmatrix} -3 \\ 9 \end{pmatrix}$ and $\overrightarrow{PN} = \begin{pmatrix} -1 \\ 3 \end{pmatrix}$ so $\overrightarrow{MP} = 3\overrightarrow{PN}$ etc.

Ratio
- Deducing the correct ratio gains you this final mark. Note 1:3 will not gain this mark.

HMSG: p 37

Q2. (a)

$y = x^3 - 3x^2 + 4$

$\Rightarrow \dfrac{dy}{dx} = 3x^2 - 6x$ ✓ ✓

For stationary points set $\dfrac{dy}{dx} = 0$ ✓

$\Rightarrow 3x^2 - 6x = 0$

$\Rightarrow 3x(x - 2) = 0 \Rightarrow x = 0$ or 2 ✓

$$x: \quad 0 \quad\quad 2$$

$\dfrac{dy}{dx} = 3x(x - 2):$

$\quad\quad + \quad - \quad +$ ✓

Shape of graph: $\quad / \quad \diagdown \quad /$

$\quad\quad$ Nature: $\quad$ max $\quad$ min

when $x = 0$ $y = 0^3 - 3 \times 0^2 + 4 = 4$
So $(0, 4)$ is a maximum ✓
stationary point.
When $x = 2$ $y = 2^3 - 3 \times 2^2 + 4 = 0$
So $(2, 0)$ is a minimum ✓
stationary point.

7 marks

Differentiate
- 1 mark for knowing to differentiate.
- 1 mark for correctly differentiating.

Strategy
- Setting $\frac{dy}{dx} = 0$ allows you to home in on the x-coordinates of the stationary points.

Solving Equation
- The common factor here is $3x$.

Justification
- 'justify their nature' means that you need a 'nature table' as shown in the solution.

y-coordinates
- 1 mark is allocated to the calculation of the two corresponding y-coordinates (4 and 0).

Interpretation
- Clear statements should be made about the type (max or min) of each stationary point.

HMSG: pp 60–61

Q2. (b)(i)

$(x + 1)(x - 2)^2$

$= (x + 1)(x - 2)(x - 2)$

$= (x + 1)(x^2 - 4x + 4)$

$= x^3 - 4x^2 + 4x + x^2 - 4x + 4$ ✓

$= x^3 - 3x^2 + 4$

Expand
- Since the answer $x^3 - 3x^2 + 4$ is given it is important that you show clearly all your steps in the expansion of the brackets.
- Notice for $(x + 1)(x^2 - 4x + 4)$ you have: $x(x^2 - 4x + 4)$ and $1(x^2 - 4x + 4)$.

1 mark

Q2. (b)(ii)

For x-intercepts set

$y = 0$

So $(x + 1)(x - 2)^2 = 0$

$\Rightarrow x + 1 = 0$ or $x - 2 = 0$

$\Rightarrow x = -1$ or $x = 2$.

Intercepts are $(-1, 0)$, $(2, 0)$

For y-intercepts set $x = 0$ ✓

So $y = 0^3 - 3 \times 0^2 + 4 = 4$ Intercept is $(0, 4)$ ✓

Sketch: ✓

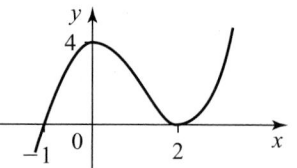

x-intercepts
- You should write down the coordinates of both x-axis intercepts, i.e. $(-1, 0)$ and $(2, 0)$.

y-intercept
- The y-axis intercept is expected to be clearly indicated when sketching. It is good practice, as with the x-intercepts, to write down the coordinates, i.e. $(0, 4)$.

Sketch
- Your sketch should clearly show the main features: shape and intercepts and stationary points.

HMSG: p 62

3 marks

Q3. (a)

$M\left(\dfrac{-3 + (-3)}{2}, \dfrac{4 + (-4)}{2}\right) = M(-3, 0)$ ✓

For $M(-3, 0)$ and $R(5, 12)$

$M_{MR} = \dfrac{12 - 0}{5 - (-3)} = \dfrac{12}{8} = \dfrac{3}{2}$ ✓

gradient of median is $\dfrac{3}{2}$

point on median is $(-3, 0)$

so equation is $y - 0 = \dfrac{3}{2}(x - (-3))$

$\Rightarrow 2y = 3(x + 3) \Rightarrow 2y = 3x + 9$

$\Rightarrow 2y - 3x = 9$ ✓

Interpretation
- Did you know what a median is? The line from a vertex to the midpoint of the opposite side.

Gradient
- The formula used here is: $A(x_1, y_1), B(x_2, y_2)\ m_{AB} = \dfrac{y_2 - y_1}{x_2 - x_1}$
- Perpendicular gradients are not required for medians, only for altitudes.

Equation
- Using $y - b = m(x - a)$ with $m = \dfrac{3}{2}$ and (a, b) being $(-3, 0)$

HMSG: p 80–81

3 marks

Q3. (b)

For $P(-3, -4)$ and $R(5, 12)$ ✓

$$m_{PR} = \frac{12 - (-4)}{5 - (-3)} = \frac{16}{8} = 2$$

$$\Rightarrow m_\perp = -\frac{1}{2}$$ ✓

gradient of altitude is $-\frac{1}{2}$

point on altitude is $Q(-3, 4)$

equation is $y - 4 = -\frac{1}{2}(x - (-3))$

$\Rightarrow 2y - 8 = -(x + 3)$
$\Rightarrow 2y - 8 = -x - 3$
$\Rightarrow 2y + x = 5$ ✓

3 marks

Strategy
- Finding the gradient of the 'base' PR is the essential 1st step here.

Perpendicular gradient
- Did you know what an altitude is? A line from a vertex perpendicular to the opposite side.
- The result used is $m = \frac{a}{b} \Rightarrow m_\perp = -\frac{b}{a}$ invert and change the sign.

Equation
- Using $y - b = m(x - a)$ with $m = -\frac{1}{2}$ and (a, b) being the point $(-3, 4)$.

HMSG: pp 78–79, p 81

Q3. (c)

To find the point of intersection:

Solve $\left.\begin{array}{c} 2y + x = 5 \\ 2y - 3x = 9 \end{array}\right\}$ ✓

Subtract: $4x = -4 \Rightarrow x = -1$ ✓

Substitute $x = -1$ in $2y + x = 5$
$\Rightarrow 2y - 1 = 5 \Rightarrow 2y = 6$
$\Rightarrow y = 3$

The point of intersection is
$S(-1, 3)$ ✓

3 marks

Strategy
- Simultaneous equations: if your method is clear you will gain this strategy mark.

Calculation
- Correct calculation of either x or y gains the 2nd mark.

Calculation
- Correct calculation of the other variable gains you this last mark.

HMSG: p 81

Q3. (d)

From (b) $m_{PR} = 2$

For $S(-1, 3)$ and $T(2, 9)$

$$m_{ST} = \frac{9 - 3}{2 - (-1)} = \frac{6}{3} = 2$$ ✓

So $m_{ST} = m_{PR} = 2$
and so ST is parallel to PR. ✓

2 marks

Gradient
- You must find the gradient of this new line ST if you are to compare its slope with that of line PR.

Parallel lines
- If the gradients of two lines are equal then the lines are parallel.
- Clear statements are needed here so that it is obvious you understand the result:
  equal gradients $\Leftrightarrow$ parallel lines.

HMSG: p 78

Q4. (a)

$$PQ = 4 \text{ units}$$

$$QR = 5 \text{ units}$$

$$RV = 6 \text{ units} \qquad \checkmark$$

1 mark

Interpretation

- The difficulty is that the axes are not shown. Look for a single change in coordinates: $P(-1, -1, -1)$ to $S(-1, 4, -1)$: This is 5 units parallel to the y-axis, since only the y-coordinate has changed.

HMSG: p 34

Q4. (b)

V(3, 4, 5)

P(−1, −1, −1)

S(−1, 4, −1)

$$\overrightarrow{VS} = \boldsymbol{s} - \boldsymbol{v}$$

$$= \begin{pmatrix} -1 \\ 4 \\ -1 \end{pmatrix} - \begin{pmatrix} 3 \\ 4 \\ 5 \end{pmatrix} = \begin{pmatrix} -4 \\ 0 \\ -6 \end{pmatrix} \qquad \checkmark$$

$$\overrightarrow{VP} = \boldsymbol{p} - \boldsymbol{v} = \begin{pmatrix} -1 \\ -1 \\ -1 \end{pmatrix} - \begin{pmatrix} 3 \\ 4 \\ 5 \end{pmatrix} = \begin{pmatrix} -4 \\ -5 \\ -6 \end{pmatrix} \qquad \checkmark$$

$$\overrightarrow{VS}.\overrightarrow{VP} = \begin{pmatrix} -4 \\ 0 \\ -6 \end{pmatrix}.\begin{pmatrix} -4 \\ -5 \\ -6 \end{pmatrix}$$

$$= -4 \times (-4) + 0 \times (-5) + (-6) \times (-6)$$

$$= 52$$

$$|\overrightarrow{VS}| = \left| \begin{smallmatrix} -4 \\ 0 \\ -6 \end{smallmatrix} \right| = \sqrt{(-4)^2 + 0^2 + (-6)^2} \qquad \checkmark$$

$$= \sqrt{16 + 0 + 36} = \sqrt{52} \qquad \checkmark$$

$$|\overrightarrow{VP}| = \left| \begin{smallmatrix} -4 \\ -5 \\ -6 \end{smallmatrix} \right| = \sqrt{(-4)^2 + (-5)^2 + (-6)^2}$$

$$= \sqrt{16 + 25 + 36} = \sqrt{77} \qquad \checkmark$$

So $\cos P\hat{V}S = \dfrac{\overrightarrow{VS}.\overrightarrow{VP}}{|\overrightarrow{VS}||\overrightarrow{VP}|} = \dfrac{52}{\sqrt{52}\sqrt{77}} \qquad \checkmark$

$$\Rightarrow P\hat{V}S = \cos^{-1}\left(\frac{52}{\sqrt{52}\sqrt{77}}\right) = 34 \cdot 73 \ldots °$$

Angle $PVS = 34 \cdot 7°$ (to 1 decimal place) $\qquad \checkmark$

7 marks

Components

- The result $\overrightarrow{AB} = \boldsymbol{b} - \boldsymbol{a}$, where $\boldsymbol{a}$ is the position vector of **A** and $\boldsymbol{b}$ is the position vector of **B**, is used frequently to find components.
- Notice that the vector arrows point outwards from the vertex of the angle. You must always do this when calculating the angle between vectors.

Dot product

- The result is $\begin{pmatrix} x_1 \\ y_1 \\ z_1 \end{pmatrix}.\begin{pmatrix} x_2 \\ y_2 \\ z_2 \end{pmatrix} = x_1x_2 + y_1y_2 + z_1z_2$

Lengths

- The result used is: $\left| \begin{smallmatrix} a \\ b \\ c \end{smallmatrix} \right| = \sqrt{a^2 + b^2 + c^2}$
- Careful with negatives. Squaring always produces a positive or zero quantity – never a negative quantity.

Angle formula

- The result used is: $\cos\theta° = \dfrac{\boldsymbol{a}.\boldsymbol{b}}{|\boldsymbol{a}||\boldsymbol{b}|}$

Calculation

- On the calculator:
$$\cos^{-1}\left(52 \div (\sqrt{52} \times \sqrt{77})\right)$$
- Radians acceptable: $0 \cdot 606$

HMSG: p 38

Q5. (a)

Upland: $u_{n+1} = 0.35u_n + 2000$ ✓

Lowland:

$$v_{n+1} = 0.25\,v_n + 2500$$ ✓

2 marks

Recurrence Relations
- Remember to use the 'percentage left' in the recurrence relation, e.g. 65% is lost so 35% remains – so use 0·35 not 0·65.

HMSG: p 91

Q5. (b)

Both recurrence relations have a limit since their multipliers (0·35 and 0·25) lie between -1 and 1. ✓

For Upland let the limit be L. ✓

Then $L = 0.35L + 2000$

$\Rightarrow L - 0.35L = 2000$

$\Rightarrow 0.65L = 2000$

$\Rightarrow L = \dfrac{2000}{0.65} = 3076.9...$ ✓

For Lowland let the limit be M.

Then $M = 0.25M + 2500$

$\Rightarrow M - 0.25M = 2500$

$\Rightarrow 0.75M = 2500$

$\Rightarrow M = \dfrac{2500}{0.75} = 3333.3...$

In the long run the Upland site requires the smaller tank (3077 litres) compared to the Lowland site (3334 litres). ✓

4 marks

Limit Condition
- 1 mark is allocated to a clear statement justifying the use of a limit. There is a limit only if the multiplier lies between -1 and 1

Strategy
- To find the limit: if you apply the recurrence relation you produce the same result, e.g. $L = 0.35L + 2000$

Limit
- Calculation of a correct limit gains 1 mark.

Decision
- For the final mark, you must correctly calculate the other limit AND clearly make your decision based on a comparison.

HMSG: pp 90–91

Q6. (a)

$P(h) = P_0 e^{-kh}$

In this case $P(4.95) = \dfrac{1}{2}P_0$ ✓

$\Rightarrow P_0 e^{-k \times 4.95} = \dfrac{1}{2}P_0 \Rightarrow e^{-4.95k} = \dfrac{1}{2}$

$\Rightarrow \log_e \dfrac{1}{2} = -4.95k \Rightarrow k = \dfrac{\log_e \frac{1}{2}}{-4.95}$ ✓

$\Rightarrow k = 0.1400$ (to 4 decimal places) ✓

3 marks

Interpretation
- You are finding k given that you know $P(h)$ and you know h

$$P(h) = P_0 e^{-kh}$$

This is $\frac{1}{2}P_0$ This is 4·95

Log statement
- Using the result: $b^c = a \leftrightarrow \log_b a = c$
  $e^{-4.95k} = \frac{1}{2}$ becomes $\log_e \frac{1}{2} = -4.95k$

Calculation
- Remember the $\boxed{\ln}$ button for 'log$_e$'.

HMSG: p 22

Q6. (b)

The formula is $P(h) = P_0 e^{-0.14h}$

For top of Everest $h = 8.85$

$P(8.85) = P_0 e^{-0.14 \times 8.85}$

$\qquad = P_0 \times 0.2896....$ ✓

This is a reduction of

$100\% - 28.96...\%$

$= 71.03...\% \doteqdot 71\%$ ✓

2 marks

Substitution

- Careful with the units. In the original formula h is in kilometres so 8·85 is used, not 8850.

Calculation

- $\boxed{e^x}$ button is used for $e^{-0.14 \times 8.85}$ so:

$$\boxed{e^x}\boxed{(}\boxed{(}\boxed{(-)}\boxed{0}\boxed{.}\boxed{1}\boxed{4}\boxed{\times}\boxed{8}\boxed{.}\boxed{8}\boxed{5}\boxed{)}\boxed{\text{EXE}}$$

or the equivalent on your calculator!

HMSG: p 22

Q7. (a) To find the points of intersection

Solve: $\left.\begin{array}{l} y = x + 2 \\ y = 6 + 4x - x^2 \end{array}\right\}$

$\qquad$ so $x + 2 = 6 + 4x - x^2$ ✓

$\qquad \Rightarrow x^2 - 3x - 4 = 0$

$\qquad \Rightarrow (x + 1)(x - 4) = 0$

$\qquad \Rightarrow x = -1$ or $x = 4$ ✓

Area $=$

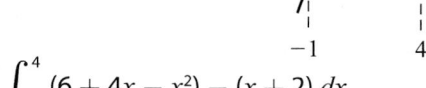

$$\int_{-1}^{4} (6 + 4x - x^2) - (x + 2)\, dx$$ ✓

$$= \int_{-1}^{4} 6 + 4x - x^2 - x - 2\, dx$$ ✓

$$= \int_{-1}^{4} 4 + 3x - x^2 dx = \left[4x + \frac{3x^2}{2} - \frac{x^3}{3} \right]_{-1}^{4}$$ ✓

$$= \left(4 \times 4 + \frac{3 \times 4^2}{2} - \frac{4^3}{3} \right)$$

$$- \left(4 \times (-1) + \frac{3 \times (-1)^2}{2} - \frac{(-1)^3}{3} \right)$$

$$= 16 + 24 - \frac{64}{3} + 4 - \frac{3}{2} - \frac{1}{3}$$ ✓

$$= 44 - \frac{65}{3} - \frac{3}{2} = \frac{264}{6} - \frac{130}{6} - \frac{9}{6}$$

$$= \frac{264 - 130 - 9}{6} = \frac{125}{6}$$

Enclosed area $= \dfrac{125}{6}$ unit^2 ✓

7 marks

Strategy

- To find the area you will be integrating, you need to know the limits. Hence, the need to set the equations equal to each other to find the points of intersection. The x values for these points are the limits used in the integration.

Solve Equation

- Never attempt to arrange a quadratic equation with a negative coefficient for the x^2 term – the factorising is then more difficult.

Strategy

- This mark is for knowing how to find the enclosed area:
$\int$(top curve) minus (bottom curve)

Limits

- Work left to right on diagram (-1 to 4) and bottom to top on integral: $\int_{-1}^{4}$

Integration

- Here you are using $\int ax^n\, dx = \dfrac{ax^{n+1}}{n+1}$.
No constant is needed when there are limits on the integral sign.

Substitution

- Careful with the order:

$\left(\underset{\text{Substitution}}{x = 4} \right) - \left(\underset{\text{Substitution}}{x = -1} \right)$ using the result

$\int_a^b f(x)dx = F(b) - F(a)$ where $F(x)$ is the result of integrating $f(x)$.

Calculation

- Take great care, even with a calculator, as these are not easy calculations!
- Always give exact answers.

HMSG: pp 94–95

Q7. (b)

$$A(x) = -\frac{5}{2}x^2 + \frac{15}{2}x + 10$$

$$\Rightarrow A'(x) = -5x + \frac{15}{2}$$

For stationary value set $A'(x) = 0$ ✓

$$\Rightarrow -5x + \frac{15}{2} = 0 \Rightarrow -5x = -\frac{15}{2} \Rightarrow x = \frac{3}{2}.$$

x:		$\frac{3}{2}$	
$A'(x)$:	$+$		$-$

✓

Shape of graph:

nature: / max \ ✓

so $x = \dfrac{3}{2}$ gives a maximum value

$$A\left(\frac{3}{2}\right) = -\frac{5}{2} \times \left(\frac{3}{2}\right)^2 + \frac{15}{2} \times \frac{3}{2} + 10$$

$$= -\frac{45}{8} + \frac{45}{4} + 10 = \frac{-45}{8} + \frac{90}{8} + \frac{80}{8}$$

$$= \frac{-45 + 90 + 80}{8} = \frac{125}{8}$$

Required fraction $= \dfrac{\frac{125}{8}}{\frac{125}{6}} = \dfrac{6}{8} = \dfrac{3}{4}.$ ✓

4 marks

Strategy

- For a maximum value you must find a stationary point where the gradient on the graph is zero, hence set $A'(x) = 0$.

Differentiate and solve

- $-5x(\times 2) = -\dfrac{15}{2}(\times 2) \Rightarrow -10x = -15$

$$\Rightarrow x = \frac{-15}{-10}$$

So $x = \frac{3}{2}$ is one method of solving the equation

Justify

- You must show $x = \frac{3}{2}$ gives a maximum value hence the need for the 'nature table'.

Interpretation

- You have found that when $x = \frac{3}{2}$ the area of the shaded triangle is at a maximum. The actual shaded area is given by $A\left(\frac{3}{2}\right)$.

  Your calculation should give $\frac{125}{8}$ unit^2 for this area. In part (a) the whole enclosed area was found to be $\frac{125}{6}$ unit^2. To calculate the fraction consider a simpler case:
  What fraction is 2 unit^2 of 6 unit^2? It is $\frac{2}{6}$ or $\frac{1}{3}$
  What fraction is
  $\frac{125}{8}$ unit^2 of $\frac{125}{6}$ unit^2? It is $\dfrac{\frac{125}{8}}{\frac{125}{6}}$

HMSG: pp 92–93

Q8.

In triangle PQR

$RQ^2 = PR^2 + PQ^2$

$\qquad = 5^2 + 12^2 = 25 + 144 = 169$

So $RQ = \sqrt{169} = 13$

$\Rightarrow \cos 2\theta = \dfrac{5}{13}$ ✓✓

$\Rightarrow 2\cos^2\theta - 1 = \dfrac{5}{13}$ ✓

$\Rightarrow 2\cos^2\theta = \dfrac{5}{13} + 1 = \dfrac{18}{13}$

$\Rightarrow \cos^2\theta = \dfrac{9}{13}$

$\Rightarrow \cos\theta = \pm\sqrt{\dfrac{9}{13}}$

$\qquad = \pm\dfrac{3}{\sqrt{13}}$ but $0 < \theta < \dfrac{\pi}{2}$ ✓

So $\cos\theta = \dfrac{3}{\sqrt{13}} = \dfrac{3 \times \sqrt{13}}{\sqrt{13} \times \sqrt{13}}$ ✓

$\qquad = \dfrac{3\sqrt{13}}{13}.$

5 marks

Strategy
- You are using SOHCAHTOA in the large right-angled triangle PQR with angle 2θ.

Value
- Pythagoras' Theorem allows you to calculate the length of the hypotenuse RQ then give the value of $\cos 2\theta$ as $\dfrac{5}{13}$

Strategy
- The Double angle formula allows you to calculate the value of $\cos\theta$.

Value
- 1st quadrant so the value of $\cos\theta$ is positive $\dfrac{3}{\sqrt{13}}$.

Rationalisation
- Rationalising the denominator – show this clearly.

HMSG: pp 54–55

Exam B: Paper 1

Q1.

$$\int 2x^2 + 3\,dx$$ ✓

$$= \frac{2x^3}{3} + 3x + C$$ ✓ ✓

3 marks

1st term
- Use of the rule: $\int x^n\,dx = \dfrac{x^n}{n+1} + \ldots$

2nd term
- Knowing how to integrate a constant:
$$\int a\,dx = ax + \ldots$$

Constant of integration
- Don't forget the constant C.

HMSG: p 68

Q2. (a)

$$f(x) = \frac{1}{1+x}$$

let $f(a) = \dfrac{1}{1+a} = b$

$$\Rightarrow 1 = b(1+a) \Rightarrow \frac{1}{b} = 1+a \Rightarrow \frac{1}{b} - 1 = a \;✓$$

so $f^{-1}(b) = \dfrac{1}{b} - 1 = a$

$$\Rightarrow f^{-1}(x) = \frac{1}{x} - 1$$ ✓

2 marks

Strategy
- If a goes in to f and b comes out, then doing this in reverse means:
  b goes in to f^{-1} and a comes out.
  In terms of the formulae you have:
  $$b = \frac{1}{1+a} \text{ with } b \text{ the subject (when } a \text{ goes in)}$$
  $$a = \frac{1}{b} - 1 \text{ with } a \text{ now the subject (} b \text{ goes in)}$$
  The strategy is to change the subject of the formula.

Formula
- Remember to use x for the final formula

Q2. (b)
The non-zero real numbers ✓

1 mark

Domain
- Division by zero in not allowed so $x \neq 0$

HMSG: p 10

Q3. (a)
Here is the sketch of $y = 2\log_3(x+3)$:

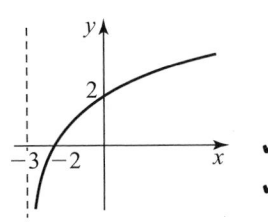

✓ ✓

2 marks

Horizontal translation
- $x+3$ indicates a left shift of three units so $(1,0)$ ends up at $(-2,0)$

Vertical scaling
- $2 \times$ at the front of the formula indicates a y-axis scaling by a factor of 2. All heights of points on the graph are doubled, so $(0,1)$ ends up at $(0,2)$ [$(3,1)$ went to $(0,1)$ by the left shift]
- The new graph approaches $x = -3$. The original graph approached the y-axis.

Q3. (b)

$$y = \log_3 \frac{1}{(x+3)^2}$$

$$= \log_3(x+3)^{-2}$$

$$= -2\log_3(x+3)$$

Here is the graph: ✓

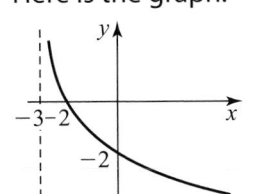

✓

2 marks

Rearrangement
- You need to rearrange the given formula using the log rules, so that the previous formula $2\log_3(x+3)$ appears.

Related graph
- The previous formula $2\log_3(x+3)$ has been multiplied by -1. This indicates you should flip the graph in the x-axis.

HMSG: pp 19–21

Q4. (a)

(i) $p.p = |p| \times |p| \times \cos 0 = 1 \times 1 \times 1 = 1$ ✓

(ii) $q.q = |q| \times |q| \times \cos 0 = \sqrt{2} \times \sqrt{2} \times 1 = 2$ ✓

(iii) $p.q = |p| \times |q| \times \cos 45° = 1 \times \sqrt{2} \times \dfrac{1}{\sqrt{2}} = 1$ ✓

3 marks

Dot product
- You are using the result: $a.b = |a|\,|b|\cos\theta$ where θ is the angle between a and b
- $|p| = 1$ since the edge length of the cube is 1 unit

Dot product
- The angle between a vector and itself is 0°
- The diagonal of the square face of the cube is at 45° to the edges of the cube.

Dot product
- You should know the exact value of cos 45°

Strategy
- Evidence of an attempt to expand the brackets: $(2p + q).(2p + q)$

Q4. (b)

$r.r = (2p + q).(2p + q)$ ✓
$= 4p.p + 4p.q + q.q$ ✓
$= 4 \times 1 + 4 \times 1 + 2 = 10$ ✓

3 marks

Expansion
- The dot product behaves very much like normal multiplication when you expand the brackets. However remember p^2 makes no sense, so don't write it!

Calculation
- This calculation uses your calculated values from part (a) of the question.

Q4. (c)

$r.r = |r|^2 = 10 \Rightarrow |r| = \sqrt{10}$ ✓

1 mark

Calculation
- $r.r = |r||r|\cos 0° = |r| \times |r| \times 1 = |r|^2$

HMSG: p 40

Q5. (a)

$\cos(A + \frac{5\pi}{6}) - \sin(A + \frac{4\pi}{3})$ ✓

$= \cos A \cos \frac{5\pi}{6} - \sin A \sin \frac{5\pi}{6} -$ ✓
$\quad (\sin A \cos \frac{4\pi}{3} + \cos A \sin \frac{4\pi}{3})$

$= \cos A \times (-\frac{\sqrt{3}}{2}) - \sin A \times \frac{1}{2} -$
$\quad \sin A \times (-\frac{1}{2}) - \cos A \times (-\frac{\sqrt{3}}{2})$

$= -\frac{\sqrt{3}}{2} \cos A - \frac{1}{2} \sin A + \frac{1}{2} \sin A + \frac{\sqrt{3}}{2} \cos A = 0$ ✓

3 marks

Expansions
- Both addition formulae expansions are given to you in your exam:
$$\sin(A \pm B) = \sin A \cos B \pm \cos A \sin B$$
$$\cos(A \pm B) = \cos A \cos B \mp \sin A \sin B$$
- Be careful to subtract the whole of the sine expansion: $-(\sin A \cos \frac{4\pi}{3} + \cos A \sin \frac{4\pi}{3})$

Exact values
- $\frac{5\pi}{6} = \pi - \frac{\pi}{6}$ a 2nd quadrant angle
- $\frac{4\pi}{3} = \pi + \frac{\pi}{3}$ a 3rd quadrant angle

Simplification
- All the terms cancel. Be very careful with the negative terms and the subtraction.

Q5. (b)

$\cos(A + \frac{5\pi}{6}) - \sin(A + \frac{4\pi}{3}) = 0$

$\Rightarrow \cos(A + \frac{5\pi}{6}) = \sin(A + \frac{4\pi}{3})$

Now let $A = \frac{\pi}{12}$

$\Rightarrow \cos(\frac{\pi}{12} + \frac{5\pi}{6}) = \sin(\frac{\pi}{12} + \frac{4\pi}{3})$

$\Rightarrow \cos(\frac{\pi}{12} + \frac{10\pi}{12}) = \sin(\frac{\pi}{12} + \frac{16\pi}{12})$

$\Rightarrow \cos \frac{11\pi}{12} = \sin \frac{17\pi}{12}$ ✓

Proof
- As with all 'show that …' questions you must be careful to show all the steps clearly.

HMSG: p 24, p 30

1 mark

Q6. (a)

$$p + q = \begin{pmatrix} -1 \\ 2 \\ 3 \end{pmatrix} + \begin{pmatrix} 1 \\ 3 \\ 2 \end{pmatrix} = \begin{pmatrix} 0 \\ 5 \\ 5 \end{pmatrix}$$

and $p - q = \begin{pmatrix} -1 \\ 2 \\ 3 \end{pmatrix} - \begin{pmatrix} 1 \\ 3 \\ 2 \end{pmatrix} = \begin{pmatrix} -2 \\ -1 \\ 1 \end{pmatrix}$ ✓

Calculations
- Add/subtract the corresponding components.

1 mark

Q6. (b)

$$(p + q) \cdot (p - q) = \begin{pmatrix} 0 \\ 5 \\ 5 \end{pmatrix} \cdot \begin{pmatrix} -2 \\ -1 \\ 1 \end{pmatrix}$$ ✓

$= 0 \times (-2) + 5 \times (-1) + 5 \times 1$
$= 0 + (-5) + 5 = 0$
$\Rightarrow p + q$ is perpendicular to $p - q$ ✓

Strategy
- Evidence that you know:
  dot product $= 0 \Rightarrow$ vectors are perpendicular

Calculations & statement
- Show the calculation clearly – working must be shown to gain this mark.
- Make a clear statement about the vectors

HMSG: p 35, pp 38–39

2 marks

Q7. (a)

$T_t = T_0 \times 10^{-kt}$
with $T_0 = 3000$, $t = 20$ and $T_t = 300$
$\Rightarrow 300 = 3000 \times 10^{-k \times 20}$ ✓
$\Rightarrow \frac{300}{3000} = 10^{-20k} \Rightarrow \frac{1}{10} = 10^{-20k} \Rightarrow 10^{-1} = 10^{-20k}$
$\Rightarrow -1 = -20k \Rightarrow k = \frac{-1}{-20} = \frac{1}{20}$ ✓

Substitution
- Correct substitution of the values into the formula will gain you this mark.

Calculation
- Notice that $a^m = a^n \Rightarrow m = n$. If the base is the same (a) then the indices (m and n) are equal.

2 marks

Q7. (b)

$T_t = T_0 \times 10^{-\frac{1}{20}t}$
Use $T_0 = 300$ and $t = 20$ ✓

giving: $T_t = 300 \times 10^{-\frac{1}{20} \times 20} = 300 \times 10^{-1} = 30$
The temperature will be 30°C. ✓

Substitution
- Alternatively you could have used:
  $T_0 = 3000$ and $t = 40$

Calculation
- Answer with a clear statement of your result.

HMSG: pp 16–17, p 22

2 marks

Q8. (a)

$$\begin{array}{r|rrrr} -4 & 1 & 6 & 0 & -32 \\ & & -4 & -8 & 32 \\ \hline & 1 & 2 & -8 & 0 \end{array}$$ ✓ ✓

Since the remainder is 0 then $x + 4$ is a factor of $x^3 + 6x^2 - 32$. ✓

Strategy
- Use of synthetic division will be expected.

Calculation
- Make sure you line up the numbers in neat columns and rows for easy reading.

Statement
- You must make the statement: 'Since the remainder is 0 then $x + 4$ is a factor'

3 marks

Q8. (b)

So $x^3 + 6x^2 - 32 = (x+4)(x^2 + 2x - 8)$ ✓

$\qquad\qquad\qquad\;\; = (x+4)(x+4)(x-2)$ ✓

$\qquad\qquad\qquad\;\; = (x+4)^2(x-2)$

2 marks

1st factorisation
- This mark is for the factor $x^2 + 2x - 8$

Full factorisation
- This is for factorising the quadratic.

HMSG: p 52

Q9. (a)

$\displaystyle\int \sqrt{1+3x}\,dx = \int (1+3x)^{\frac{1}{2}}\,dx$ ✓

$\displaystyle\qquad = \frac{(1+3x)^{\frac{3}{2}}}{\frac{3}{2} \times 3} + C$ ✓

$\displaystyle\qquad = \frac{2(1+3x)^{\frac{3}{2}}}{9} + C = \frac{2\left(\sqrt{(1+3x)}\right)^3}{9} + C$ ✓

3 marks

Preparation
- This involves changing the expression into a form suitable for integration – in this case changing the root to the power of $\frac{1}{2}$

Integration
- Use of $\displaystyle\int (ax+b)^n\,dx = \frac{(ax+b)^{n+1}}{a(n+1)} + ...$

Simplification & constant
- To simplify: multiply top and bottom of the fraction by 2
- The last step is not needed to gain this mark, however, it is needed for the following question.

Q9. (b)

$\displaystyle\int_0^1 \sqrt{1+3x}\,dx = \left[\frac{2\left(\sqrt{(1+3x)}\right)^3}{9}\right]_0^1$

$\displaystyle= \frac{2\left(\sqrt{(1+3\times1)}\right)^3}{9} - \frac{2\left(\sqrt{(1+3\times0)}\right)^3}{9}$ ✓

$\displaystyle= \frac{2\left(\sqrt{4}\right)^3}{9} - \frac{2\left(\sqrt{1}\right)^3}{9} = \frac{2\times8}{9} - \frac{2\times1}{9}$

$\displaystyle\qquad\qquad = \frac{16}{9} - \frac{2}{9} = \frac{14}{9}$ ✓

2 marks

Substitution
- To proceed with substitution you need to know the meaning or $a^{\frac{m}{n}}$: take the nth root then raise to the power m.

Calculation
- Correct evaluation gains you this mark.

HMSG: pp 70–71

Q10. (a)

$u_1 = pu_0 - 1 = p \times 3 - 1 = 3p - 1$ ✓

$u_2 = pu_1 - 1 = p(3p-1) - 1 = 3p^2 - p - 1$ ✓

2 marks

1st term
- 'in terms of p' so you will not get a numerical value. You require an expression involving p.

2nd term
- u_1 is replaced by the expression $3p - 1$ and you will be expected to simplify the result.

Q10. (b)

$u_2 = 1 \Rightarrow 3p^2 - p - 1 = 1 \Rightarrow 3p^2 - p - 2 = 0$ ✓

$\Rightarrow (3p+2)(p-1) = 0$

$\Rightarrow 3p+2 = 0 \text{ or } p - 1 = 0 \Rightarrow p = -\frac{2}{3} \text{ or } p = 1$ ✓

Since p is the multiplier and there is only a limit if the multiplier lies between -1 and 1, then $p = -\frac{2}{3}$ is the only value generating a sequence with a limit. ✓

3 marks

Equation
- This is a quadratic equation in the variable p.

Roots
- Solving the quadratic equation and stating the two roots will gain you this mark.

Solution & Reason
- For $u_{n+1} = pu_n - 1$ to have a limit then $-1 < p < 1$. This must appear somewhere in your reasoning when you eliminate $p = 1$.

Q10. (c)

Let the limit $= L$ then since $u_{n+1} = -\frac{2}{3}u_n - 1$
this gives: $L = -\frac{2}{3}L - 1$ ✓
$\Rightarrow 3L = -2L - 3 \Rightarrow 5L = -3 \Rightarrow L = -\frac{3}{5}$
The limit is $-\frac{3}{5}$ ✓

Strategy
- If you use $L = \dfrac{c}{1-m}$ then use that method.

Calculation
- Multiply through by 3 to remove the fraction.

HMSG: pp 88–90

2 marks

Q11.

$$y = \frac{1}{16}x^4 - \frac{1}{8}x^2 + x \quad ✓$$

$$\Rightarrow \frac{dy}{dx} = \frac{1}{4}x^3 - \frac{1}{4}x + 1 \quad ✓$$

The tangent $y = x + c$ has gradient 1
So set
$$\frac{dy}{dx} = 1 \quad ✓✓$$

$$\Rightarrow \frac{1}{4}x^3 - \frac{1}{4}x + 1 = 1$$

$$\Rightarrow \frac{1}{4}x^3 - \frac{1}{4}x = 0 \Rightarrow x^3 - x = 0$$

$$x(x^2 - 1) = 0 \Rightarrow x(x-1)(x+1) = 0$$

$$\Rightarrow x = 0 \text{ or } x = 1 \text{ or } x = -1 \quad ✓$$

For $x = 0$:
$$y = \frac{1}{16} \times 0^4 - \frac{1}{8} \times 0^2 + 0 = 0 \quad ✓$$

so $y = x + c$ gives $0 = 0 + c \Rightarrow c = 0$
The tangent is $y = x$ with contact point $(0, 0)$

For $x = -1$:
$$y = \frac{1}{16} \times (-1)^4 - \frac{1}{8} \times (-1)^2 + (-1)$$

$$= -\frac{17}{16}$$

So $y = x + c$ gives
$$-\frac{17}{16} = -1 + c \Rightarrow c = -\frac{1}{16}$$

tangent is $y = x - \frac{1}{16}$, ✓

contact point is $\left(-1, -\frac{17}{16}\right)$

For $x = 1$:
$$y = \frac{1}{16} \times 1^4 - \frac{1}{8} \times 1^2 + 1 = \frac{15}{16}$$

so $y = x + c$ gives
$$\frac{15}{16} = 1 + c \Rightarrow c = -\frac{1}{16}$$

Tangent is $y = x - \frac{1}{16}$ (same as for $x = -1$)

Contact point is $\left(1, \frac{15}{16}\right)$ ✓

Strategy
- Evidence that you know to differentiate will gain you this mark.

Differentiation
- Be careful with the fractions here:
$4 \times \frac{1}{16} = \frac{4}{16} = \frac{1}{4}$ and $2 \times \frac{1}{8} = \frac{2}{8} = \frac{1}{4}$

Gradient
- Compare $y = mx + c$ and $y = x + c$. This leads to $m = 1$ for the gradient of the tangent line.

Strategy
- This second strategy mark is given for knowing to set the gradient formula equal to 1.

Solving
- Remove fractions first by multiplying both sides of the equation by 4. Although this is a cubic, terms are missing, and so it is easily factorised once you realise to remove the common factor x.

Calculation
- This processing mark involves a fair amount of calculation and is gained for clearly stating the possible values of c. These are:
$$c = 0 \text{ and } c = -\frac{1}{16}$$

Interpretation
- There are three points of contact:
$$(0,0), \left(-1, -\frac{17}{16}\right) \text{ and } \left(1, \frac{15}{16}\right)$$

but only two equations for the tangent:
$$y = x \text{ and } y = x - \frac{1}{16}$$

A close examination of the graph shows $y = x - \frac{1}{16}$ is a tangent at two separate points on the curve as is shown in this diagram

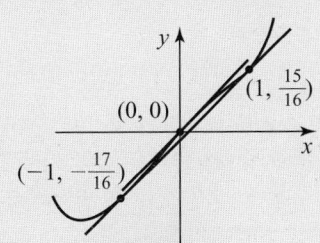

HMSG: p 60

8 marks

Q12. (a)

$f(x) = 2x - 1, g(x) = \log_{12} x$

$f(g(x)) + g(f(x)) = 0$ ✓

$\Rightarrow f(\log_{12} x) + g(2x - 1) = 0$ ✓

$\Rightarrow 2\log_{12} x - 1 + \log_{12}(2x - 1) = 0$ ✓

$\Rightarrow \log_{12} x^2 + \log_{12}(2x - 1) = 1$

$\Rightarrow \log_{12} x^2(2x - 1) = 1$ ✓

$\Rightarrow x^2(2x - 1) = 12^1$ ✓

$\Rightarrow 2x^3 - x^2 - 12 = 0$

$$
\begin{array}{r|rrrr}
2 & 2 & -1 & 0 & -12 \\
  &   & 4 & 6 & 12 \\
\hline
  & 2 & 3 & 6 & 0 \\
\end{array}
$$
✓

So $x = 2$ satisfies the equation and is therefore a solution.

6 marks

Composition
- Combining f and g to get the formula $f(g(x))$ is called the 'composition' of f and g. Showing either $f(\log_{12} x)$ or $g(2x - 1)$ gains you this first mark.

Composition
- Reaching either $2\log_{12} x - 1$ or $\log_{12}(2x - 1)$ will gain you this 2nd mark.

Composition
- Correctly finding the 2nd expression $2\log_{12} x - 1$ or $\log_{12}(2x - 1)$ will give you this 3rd mark.

Log Law
- The rule used here is:
$$\log_b m + \log_b n = \log_b mn$$

Exponential form
- The result used is:
$$\log_b a = c \;\leftrightarrow\; a = b^c$$
$$\text{(logarithmic form)} \qquad \text{(exponential form)}$$

Calculation
- At this stage you must clearly show that $x = 2$ is a solution of $2x^3 - x^2 - 12 = 0$. The table shows that $f(2) = 0$ where $f(x) = 2x^3 - x^2 - 12$.

HMSG: p 10, p 20, p 52

Q12. (b)

From part(a) equation becomes:

$(x - 2)(2x^2 + 3x + 6) = 0$

Consider $2x^2 + 3x + 6 = 0$ ✓

Discriminant $= 3^2 - 4 \times 2 \times 6$

$= 9 - 48 = -39$

Since Discriminant < 0 there are no Real solutions. ✓

The only Real solution is $x = 2$.

2 marks

Strategy
- One solution, $x = 2$, comes from the factor $x - 2$ equating to zero. Any other solutions will therefore come from equating the other factor to zero, i.e. $2x^2 + 3x + 6 = 0$. If your working shows you knew this you will gain this strategy mark.

Communication
- There must be a clear reason for 'no Real solutions'. In this case, the discriminant of the quadratic equation is negative. Your working should state this fact quite clearly. The result used is:

Discriminant $< 0 \Rightarrow$ no Real roots.

HMSG: p 46

Exam B: Paper 2

Q1. (a)

$f(x) = x^3 + 3x^2 - 4$ ✓

$\Rightarrow f'(x) = 3x^2 + 6x$

For stationary points set

$f'(x) = 0$ ✓

$\Rightarrow 3x^2 + 6x = 0 \Rightarrow 3x(x + 2) = 0$ ✓

$\Rightarrow x = 0$ or $x = -2$

x: $\quad -2 \qquad 0$ ✓

$f'(x) = 3x(x+2)$: $\quad + \quad - \quad +$

Shape of graph :

nature: $\qquad$ max $\quad$ min

$f(-2) = (-2)^3 + 3x\,(-2)^2 - 4 = 0$

So $(-2, 0)$ is a maximum stationary point

$f(0) = 0^3 + 3 \times 0^2 - 4 = -4$ ✓

So $(0, -4)$ is a minimum stationary point ✓

6 marks

Differentiate
- Correct differentiation will gain this mark.

Strategy
- To find the stationary points set $f'(x) = 0$.

Solutions
- The common factor is $3x$ with two roots: 0 and -2.

Justify
- A 'nature table' is needed to determine maximum and minimum points.

y-coordinates
- Take care with negatives. In paper 1 questions you have no access to a calculator. In general remember:
  squaring produces positive or zero answers, cubing a negative quantity gives a negative answer. In this case: $(-2)^3 = -8$ and $(-2)^2 = 4$.

Statements
- Clear statements concerning the nature of the points, i.e. maximum or minimum, etc. are expected for this final mark.

HMSG: pp 60–61

Q1. (b)

For x-intercepts set $y = 0$

$\quad$ So $x^3 + 3x^2 - 4 = 0$

$\quad \Rightarrow (x - 1)(x + 2)^2 = 0$

$\quad \Rightarrow x = 1$ or $x = -2$

Intercepts are $(-2, 0)$ and $(1, 0)$ ✓

For y-intercept set $x = 0$

So $y = 0^3 + 3 \times 0^2 - 4 = -4$

Intercept is $(0, -4)$ ✓

Sketch:

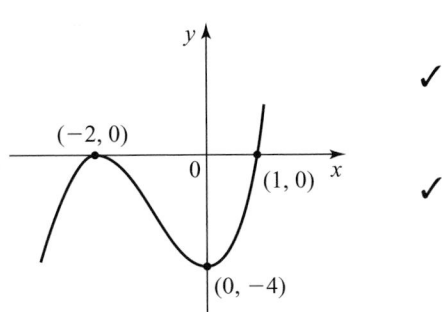

4 marks

x-intercepts
- x-axis intercepts are obtained by setting $y = 0$ or $f(x) = 0$. In this case, a cubic equation is formed, but you have been given the factorisation of the cubic. So you can proceed by setting each factor in turn equal to 0.
- Remember repeated roots, like $x = -2$ in this case, would indicate that the graph touches the x-axis

y-intercept
- 1 mark is allocated for this result.

Sketch
- 1 mark is allocated for your sketch showing the cubic shape correctly, with the two stationary points clearly shown and labelled.
- The 2nd mark is given for your sketch clearly showing the intercepts, i.e. $(0,-4)$ and $(1,0)$, etc.

HMSG: p 62

Q2. (a) (i)

$x^2 + y^2 + 4x - 6y + 5 = 0$

Centre is $(-2, 3)$. ✓

1 mark

Centre
- The result used is:
$$x^2 + y^2 + 2gx + 2fy + c = 0$$
Centre: $(-g, -f)$
This result is on your formulae sheet in the exam.

HMSG: p 82

Q2. (a) (ii)

Radius $= \sqrt{(-2)^2 + 3^2 - 5}$

$= \sqrt{4 + 9 - 5} = \sqrt{8} = 2\sqrt{2}$ ✓

1 mark

Radius
- Notice that:
$$\sqrt{8} = \sqrt{4 \times 2} = \sqrt{4} \times \sqrt{2} = 2 \times \sqrt{2} = 2\sqrt{2}$$
- The radius formula: $\sqrt{g^2 + f^2 - c}$ for the circle $x^2 + y^2 + 2gx + 2fy + c = 0$ is given to you on your formulae sheet in the exam.

HMSG: p 82

Q2. (b) (i)

Circle B:
$$(x - 2)^2 + (y + 1)^2 = 2$$
has centre $(2, -1)$ ✓

Distance between $(-2, 3)$ and $(2, -1)$ is:

$\sqrt{(-2 - 2)^2 + (3 - (-1))^2}$

$= \sqrt{(-4)^2 + 4^2} = \sqrt{32} = 4\sqrt{2}$ ✓

2 marks

Centre
The result used is:
- For circle $(x - a)^2 + (y - b)^2 = r^2$ the centre is (a, b) a result also given to you in your exam.

Calculation
- The distance formula is used here. If $A(x_1, y_1)$ and $B(x_2, y_2)$ then
$$AB = \sqrt{(x_2 - x_1)^2 + (y_2 - y_1)^2}$$

HMSG: p 80, p 82

Q2. (b) (ii)

Circle A: radius $= 2\sqrt{2}$

Circle B: radius $= \sqrt{2}$

Distance between centres $= 4\sqrt{2}$

Sum of radii $= 2\sqrt{2} + \sqrt{2} = 3\sqrt{2}$ ✓

less than $4\sqrt{2}$, the distance between centres

$\Rightarrow$ circles do not intersect ✓

2 marks

Strategy

I — Distance between centres is greater than the sum of the radii.

II — Distance between centres is equal to the sum of the two radii.

III — Distance between centres is less than the sum of the two radii.

Communication
- There must be a comparison given. In this case a statement that the sum of the radii is $3\sqrt{2}$, which is less than the centre distance $4\sqrt{2}$, so there is no intersection.

(This is situation I shown above.)

HMSG: p 86

Q2. (c)

Solve $\quad y = x + 5$

$x^2 + y^2 + 4x - 6y + 5 = 0$

Substitute $y = x + 5$ in circle equation: ✓

$x^2 + (x+5)^2 + 4x - 6(x+5) + 5 = 0$ ✓

$\Rightarrow x^2 + x^2 + 10x + 25 + 4x - 6x - 30 + 5 = 0$

$\Rightarrow 2x^2 + 8x = 0 \Rightarrow 2x(x+4) = 0$ ✓

$\Rightarrow x = 0$ or $x = -4$ ✓

when $x = -4 \qquad y = -4 + 5 = 1$

when $x = 0 \qquad y = 0 + 5 = 5$

So $P(-4, 1)$ and $Q(0, 5)$ ✓

5 marks

Strategy
- In general, to find where graphs $y = f(x)$ and $y = g(x)$ intersect then you equate the formulae i.e. $f(x) = g(x)$ and solve the resulting equation. However in this case, the equation of the circle cannot be written as '$y = ...$' so substitution is used.

Substitution
- Replace all occurrences of y by $x + 5$ in the circle equation.

'Standard form'
- Reducing the equation to $2x^2 + 8x = 0$ gains you this mark.

Solve for x
- The common factor is $2x$ with roots -4 and 0.

Coordinates
- Coordinates are asked for not just values of x and y.

HMSG: p 84

Q3. (a)

$A(-5, 2)$ and $B(-3, 8)$ ✓

$\Rightarrow m_{AB} = \dfrac{8-2}{-3-(-5)} = \dfrac{6}{2} = 3$

$\Rightarrow m_\perp = -\dfrac{1}{3}$ ✓

Midpoint of AB is

$\left(\dfrac{-5+(-3)}{2}, \dfrac{2+8}{2}\right) = (-4, 5)$ ✓

For the perpendicular bisector:

A point on the line is $(-4, 5)$

and the gradient $= -\dfrac{1}{3}$

So, equation is:

$y - 5 = -\dfrac{1}{3}(x - (-4))$

$\Rightarrow 3y - 15 = -(x + 4)$

$\Rightarrow 3y - 15 = -x - 4$ ✓

$\Rightarrow 3y + x = 11$

4 marks

Strategy
- Here you are using the gradient formula:
  $P(x_1, y_1), Q(x_2, y_2)$ gives $m_{pq} = \frac{y_2 - y_1}{x_2 - x_1}$

Perpendicular Gradient
- If $m = \frac{a}{b}$ then $m_\perp = -\frac{b}{a}$. For $m = 3$ you think of 3 as $\frac{3}{1}$. Inverting and changing sign then gives $-\frac{1}{3}$ as shown in the solution.

Strategy
- You have to know that 'bisector' means find the mid point of AB.
- The mid point formula is:
  $P(x_1, y_1), Q(x_2, y_2)$.
  Midpoint is $\left(\frac{x_1+x_2}{2}, \frac{y_1+y_2}{2}\right)$

Equation
- Using $y - b = m(x - a)$ with $m = -\frac{1}{3}$ and the point (a,b) is $(-4,5)$.

HMSG: pp 76–80

Q3. (b)

$A(-5, 2)$ and $D(5, -8)$

Midpoint of AD is:

$\left(\dfrac{-5+5}{2}, \dfrac{2+(-8)}{2}\right) = M(0, -3)$ ✓

So, using $C(3, 6)$ and $M(0, -3)$

$m_{CM} = \dfrac{6-(-3)}{3-0} = \dfrac{9}{3} = 3$ ✓

For the median:

A point on the line is $(0, -3)$ and the gradient is 3

So, equation is $y - (-3) = 3(x - 0)$

$\Rightarrow y + 3 = 3x$

$\Rightarrow y - 3x = -3$ ✓

Strategy
- The median is the line from C to the midpoint of the opposite side AD. You will therefore need to find the coordinates of the midpoint.

Gradient
- Medians do not involve 'perpendicular' and so when m_{CM} is calculated you use this value, 3, to find the equation of the median.

Equation
- Use $y - b = m(x - a)$ with $m = 3$ and (a,b) being the point $(0,-3)$. Alternatively, spot that $(0,-3)$ is the y-intercept of the line and use $y = mx + c$ with $m = -3$ and $c = -3$ to give $y = 3x - 3$.

HMSG: pp 80–81

3 marks

Q3. (c)

To find the intersection point S: ✓

Solve:

$\left.\begin{array}{l} 3y + x = 11 \\ y - 3x = -3 \end{array}\right\}_{(\times 3)}$ $\begin{array}{l} \rightarrow 3y + x = 11 \\ \rightarrow \underline{3y - 9x = -9} \end{array}$

subtract: $10x = 20$

$\Rightarrow x = 2$ ✓

Now substitute $x = 2$ in

$y - 3x = -3$

$\Rightarrow y - 3 \times 2 = -3 \Rightarrow y - 6 = -3$

$\Rightarrow y = 3$ ✓

so $S\,(2,3)$

Strategy
- To find the point of intersection of two lines you solve the two equations of the lines simultaneously.

Find one variable
- An alternative is to multiply the 1st equation by 3, then add to give $10y = 30 \Rightarrow y = 3$.

Second variable
- Use the 'easier' equation when doing the substitution. If you found $y = 3$ first then the 1st equation is 'easier'.

HMSG: p 81

3 marks

Q4.

$3\cos 2x° + 9\cos x° = \cos^2 x° - 7$

$\Rightarrow 3(2\cos^2 x° - 1) + 9\cos x° = \cos^2 x° - 7$ ✓

$\Rightarrow 6\cos^2 x° - 3 + 9\cos x° = \cos^2 x° - 7$

$\Rightarrow 5\cos^2 x° + 9\cos x° + 4 = 0$ ✓

$\Rightarrow (5\cos x° + 4)(\cos x° + 1) = 0$ ✓

$\Rightarrow 5\cos x° + 4 = 0 \text{ or } \cos x° + 1 = 0$

$\Rightarrow \cos x° = -\dfrac{4}{5} \text{ or } \cos x° = -1$ ✓

For $\cos x° = -\dfrac{4}{5}$

$x°$ is in 2nd or 3rd quadrants

1st quadrant angle is 36·9°

so $x = 180 - 36·9$ or

$\quad x = 180 + 36·9$

$\Rightarrow x = 143·1 \text{ or } x = 216·9$

For $\cos x° = -1$

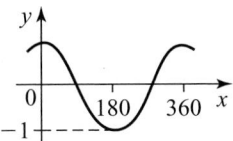

So $x = 180$

Solutions are:

143·1, 180, 216·9 ✓

(to 1 decimal place).

5 marks

Strategy

- There are three expansions of $\cos 2x°$:
  $\cos 2x° = 2\cos^2 x - 1$ or $\cos^2 x - \sin^2 x$ or
  $1 - 2\sin^2 x$. Which of these you use is dictated
  by the surrounding 'landscape' in the equation:
  There is a 'cos x term' and a cos $2x$ term,
  but no 'sin x term' or 'sin$^2 x$ term'. So
  $\cos 2x = 2\cos^2 x - 1$ is used as the other
  two forms involve $\sin^2 x$.

'Standard form'.

- You recognize the equation as a quadratic
  equation in $\cos x$ and arrange it into the
  standard form
  $5\cos^2 x° + 9\cos x° + 4 = 0$.

Factorisation

- Compare $5c^2 + 9c + 4 = (5c + 4)(c + 1)$.
- Remember that you should always check your
  answer in a factorisation by multiplying out,
  i.e. working backwards.

Solving for $\cos x$

- From $(5c + 4)(c + 1) = 0$ to $c = -\frac{4}{5}$ or
  $c = -1$ is no different to what you do at this
  stage except that the single variable c is replaced
  by the expression $\cos x°$.

Solutions

- A lot of knowledge and work needed for this
  final processing mark!
- The quadrant diagram is used:

  for $\cos x°$ negative.

HMSG: p 55

Q5. (a)

$M(0,3,2)$ ✓
$N(5,2,0)$ ✓

2 marks

Point M
- M is the midpoint. Half-way along DG which is 6 units long (y-coordinate) is 3 units.

Point N
- N is $\frac{1}{3}$ of the way along AB so $\frac{1}{3}$ of $6 = 2$ units is the y-coordinate.

HMSG: p 34, p 80

Q5. (b)

$M(0,3,2)$ and $B(5,6,0)$

$$\overrightarrow{MB} = \boldsymbol{b} - \boldsymbol{m} = \begin{pmatrix} 5 \\ 6 \\ 0 \end{pmatrix} - \begin{pmatrix} 0 \\ 3 \\ 2 \end{pmatrix} = \begin{pmatrix} 5 \\ 3 \\ -2 \end{pmatrix} \checkmark$$

also $M(0,3,2)$ and $N(5,2,0)$

$$\overrightarrow{MN} = \boldsymbol{n} - \boldsymbol{m} = \begin{pmatrix} 5 \\ 2 \\ 0 \end{pmatrix} - \begin{pmatrix} 0 \\ 3 \\ 2 \end{pmatrix} = \begin{pmatrix} 5 \\ -1 \\ -2 \end{pmatrix} \checkmark$$

2 marks

Components
- The basic result used is:

$P(x_1,y_1,z_1)$ and $Q(x_2,y_2,z_2)$

$$\text{so } \overrightarrow{PQ} = \boldsymbol{q} - \boldsymbol{p} = \begin{pmatrix} x_2 \\ y_2 \\ z_2 \end{pmatrix} - \begin{pmatrix} x_1 \\ y_1 \\ z_1 \end{pmatrix} = \begin{pmatrix} x_2 - x_1 \\ y_2 - y_1 \\ z_2 - z_1 \end{pmatrix}$$

HMSG: p 36

Q5. (c)

Use $\cos\theta = \dfrac{\boldsymbol{v}.\boldsymbol{w}}{|\boldsymbol{v}|\,|\boldsymbol{w}|}$ ✓

where $\boldsymbol{v} = \begin{pmatrix} 5 \\ 3 \\ -2 \end{pmatrix}$ and $\boldsymbol{w} = \begin{pmatrix} 5 \\ -1 \\ -2 \end{pmatrix}$

$$\boldsymbol{v}.\boldsymbol{w} = \begin{pmatrix} 5 \\ 3 \\ -2 \end{pmatrix} . \begin{pmatrix} 5 \\ -1 \\ -2 \end{pmatrix}$$

$$= 5 \times 5 + 3 \times (-1) + (-2) \times (-2)$$
$$= 25 - 3 + 4 = 26 \quad \checkmark$$

$$|\boldsymbol{v}| = \sqrt{5^2 + 3^2 + (-2)^2}$$
$$= \sqrt{25 + 9 + 4} = \sqrt{38} \quad \checkmark$$

$$|\boldsymbol{w}| = \sqrt{5^2 + (-1)^2 + (-2)^2} \quad \checkmark$$
$$= \sqrt{25 + 1 + 4} = \sqrt{30}$$

So, $\cos\theta = \dfrac{26}{\sqrt{38}\sqrt{30}}$

$$\Rightarrow \theta = \cos^{-1}\left(\frac{26}{\sqrt{38}\sqrt{30}} \right)$$

So $\theta = 39.64\ldots \doteqdot 39.6$ (to 1 dec pl.) ✓

5 marks

Strategy
- This strategy is for the use of the 'scalar' or 'dot' product formula.

Calculation
- The 'dot product' formula is:

$$\begin{pmatrix} x_1 \\ y_1 \\ z_1 \end{pmatrix} . \begin{pmatrix} x_2 \\ y_2 \\ z_2 \end{pmatrix} = x_1 x_2 + y_1 y_2 + z_1 z_2$$

Magnitudes
- The magnitude formula is:

$$\left\| \begin{pmatrix} x_1 \\ y_1 \\ z_1 \end{pmatrix} \right\| = \sqrt{x_1^2 + y_1^2 + z_1^2}$$

Angle
- Be careful with your calculator calculation

$\boxed{\text{INV}}\,\boxed{\cos}\,(\,(\,2\,6\,\div\,(\,\sqrt{\,}\,3\,8\,\times\,\sqrt{\,}\,3\,0\,)\,)\,)\,\boxed{\text{EXE}}$

The brackets are vital: $\cos^{-1}(\ldots)$ and $(\sqrt{38} \times \sqrt{30})$.

HMSG: p 38

Q6. (a)

$m = 1$ and $n = \sqrt{3}$ ✓
✓

Interpret graphs
- The 'normal' sine graph has amplitude $= 1$. This gives $m = 1$. The amplitude of the cosine graph shown is $\sqrt{3}$ so $n = \sqrt{3}$.

HMSG: pp 26–27

2 marks

Q6. (b)

$f(x) = \sin x$ and $g(x) = \sqrt{3}\,\cos x$

So $f(x) - g(x) = \sin x - \sqrt{3}\,\cos x$

Let $\sin x - \sqrt{3}\,\cos x$
$$= k\,\sin(x - a),\ k > 0$$
$$\Rightarrow \sin x - \sqrt{3}\,\cos x$$
$$= k\,\sin x\,\cos a - k\,\cos x\,\sin a \quad ✓$$

Now equate coefficients of $\sin x$ and $\cos x$:

$\left.\begin{array}{l} k\cos a = 1 \\[6pt] k\sin a = \sqrt{3} \end{array}\right\}$ since both $\sin a$ and $\cos a$ are positive a is in 1st quadrant ✓

Divide: $\dfrac{k\sin a}{k\cos a} = \dfrac{\sqrt{3}}{1}$

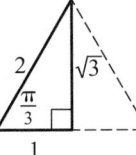

$$\Rightarrow \tan a = \sqrt{3}$$
$$\Rightarrow a = \frac{\pi}{3} \quad ✓$$

Square and add:
$$(k\cos a)^2 + (k\sin a)^2 = 1^2 + \left(\sqrt{3}\right)^2$$
$$\Rightarrow k^2\cos^2 a + k^2\sin^2 a = 1 + 3$$
$$\Rightarrow k^2(\cos^2 a + \sin^2 a) = 4$$
$$\Rightarrow k^2 \times 1 = 4 \Rightarrow k = 2\ (k > 0) \quad ✓$$

So, $f(x) - g(x) = \sin x - \sqrt{3}\,\cos x$
$$= 2\sin\left(x - \frac{\pi}{3}\right)$$

Strategy
- You must clearly show the expansion of $k\sin(x - a)$. To do this you use the addition formulae:
$$\sin(A \pm B) = \sin A \cos B \pm \cos A \sin B$$
which is given on your formulae sheet.

Compare coefficients

compare $\boxed{1}\ \sin x \qquad -\ \boxed{\sqrt{3}}\ \cos x$

$\boxed{k\sin x\cos a} \quad -\ \boxed{k\cos x\sin a}$

giving $\quad k\cos a = 1 \quad$ and $\quad k\sin a = \sqrt{3}$

Find a
- You should recognise $\sqrt{3}$ as an exact value leading to the angle $\frac{\pi}{3}$.

Find k
- For finding a you used $\frac{\sin a}{\cos a} = \tan a$ and for finding k you use $\sin^2 a + \cos^2 a = 1$. If you try to apply a learnt 'formula' for finding k sometimes mistakes creep in. It is better to understand that squaring and adding both sides of the two equations leads to the value of k, and just 'do the mathematics' at the time, $k^2 = 4$, $k = -2$ is not an allowable value since $k > 0$.

HMSG: pp 32–33

4 marks

Q6. (c)

$$y = 2\sin\left(x - \frac{\pi}{3}\right)$$

$$\Rightarrow \frac{dy}{dx} = 2\cos\left(x - \frac{\pi}{3}\right) \qquad \checkmark$$

For a gradient of 2, set $\dfrac{dy}{dx} = 2$

$$\Rightarrow 2\cos\left(x - \frac{\pi}{3}\right) = 2$$

$$\Rightarrow \cos\left(x - \frac{\pi}{3}\right) = 1$$

$$\Rightarrow x - \frac{\pi}{3} = 0 \text{ (or } 2\pi\text{)}$$

$$\Rightarrow x = \frac{\pi}{3}\left(\text{or } 2\pi + \frac{\pi}{3}\right)$$

But $0 \leq x \leq \pi$ so $x = \dfrac{\pi}{3}$
is the only solution. $\qquad \checkmark$

2 marks

Strategy
- You are told the gradient so you have to differentiate and find the x-value that satisfies $\frac{dy}{dx} = 2$. The word 'hence' is very important. It is telling you to use the previous result. You will lose marks if you do not do this.

Differentiation involves the chain rule:

$$y = 2\sin(g(x)) \Rightarrow \frac{dy}{dx} = 2\cos(g(x)) \times g'(x)$$

In this case $g(x) = x - \frac{\pi}{3}$ so $g'(x) = 1$ so

$$\frac{dy}{dx} = 2\cos\left(x - \frac{\pi}{3}\right) \times 1 = 2\cos\left(x - \frac{\pi}{3}\right)$$

Equation and solution
- For $\cos\theta = 1$

$$\theta = \ldots -2\pi, \, 0, \, 2\pi \ldots$$

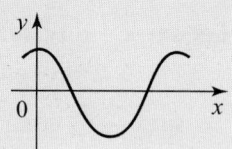

- In this question, the angle is $x - \frac{\pi}{3}$ so to finally find x you must add $\frac{\pi}{3}$.

HMSG: p 54, pp 64–65

Q7.

Intersection with $y = 15$ solve:

$$\left.\begin{array}{l} y = 15 \\ y = x^4 - 1 \end{array}\right\} \text{ so } x^4 - 1 = 15$$
$$\Rightarrow x^4 = 16 \Rightarrow x = 2 \text{ or } -2$$

For x-intercept set $y = 0 \qquad \checkmark$
so $x^4 - 1 = 0 \Rightarrow x^4 = 1$
$\qquad \Rightarrow x = 1 \text{ or } -1 \qquad \checkmark$
Here is the diagram:

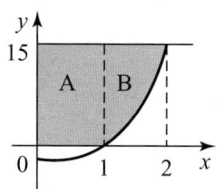

Area A is a rectangle:
$\qquad$ Area $A = 1 \times 15 = 15$ unit^2

Limits
- It is essential to find the x values of the intersection of the lines $y = 15$ and x-axis with the curve. These values will be used later in the integral for finding the area between two graphs.

x-values
- The relevant values are $x = 1$ and $x = 2$. Notice that all the work can be done in the 1st quadrant and then doubled at the end, since the y-axis is an axis of symmetry for the diagram.

(continued to next page)

Q7. Continued

Area $B =$

$$= \int_1^2 15 - (x^4 - 1) \, dx \qquad ✓$$

$$= \int_1^2 16 - x^4 \, dx = \left[16x - \frac{x^5}{5} \right]_1^2 \qquad ✓$$

$$= \left(16 \times 2 - \frac{2^5}{5} \right) - \left(16 \times 1 - \frac{1^5}{5} \right) \qquad ✓$$

$$= 32 - \frac{32}{5} - 16 + \frac{1}{5} = 16 - \frac{31}{5} = \frac{49}{5} \text{ unit}^2 ✓$$

Area A + Area B

$$= 15 + \frac{49}{5} = \frac{75}{5} + \frac{49}{5} = \frac{124}{5} \text{ unit}^2 \qquad ✓$$

By symmetry, the required area

$$= 2 \times \frac{124}{5} = \frac{248}{5} \text{ units}^2 \qquad ✓$$

8 marks

Strategy
- You should know to use integration to find the area between $y = 15$ and the curve (area B) and add the area of Rectangle A. There are other methods possible.

Integration
- Notice
$$\int a \, dx = ax + c \text{ and } \int x^n = \frac{x^{n+1}}{n+1} + c$$
(a is a constant)
- The constant c is not needed when there are limits.

Limits
- A mark is allocated for correct use of limits 1 and 2.

Evaluate
- Using $\int_a^b f(x) \, dx = [F(x)]_a^b = F(b) - F(a)$
Where $F(x)$ is the result of integrating $f(x)$.

Strategy
- Knowing what to add together!

Calculation
- Final answer is gained by doubling as you have only found the area in the 1st quadrant.

HMSG: pp 94–95

Q8. (a)

(0,9) lies on the curve
So $y = 9$ when $x = 0$
$\quad y = -x^2 + a$ gives $9 = -0^2 + a$
$\Rightarrow a = 9$. ✓

1 mark

Calculation
- If (p,q) lies on a graph with equation $y = f(x)$ then $q = f(p)$. In other words, the coordinates of the point can be substituted into the equation. In this case, this gives the value of a.

HMSG: p 9

Q8. (b)

$f(x) = -x^2 + 9 = 9 - x^2$
To find coordinates of P set $x = m$
So $f(m) = 9 - m^2 \Rightarrow P(m, 9 - m^2)$
$\quad AP = 9 - m^2$ ✓

1 mark

Calculation of y-coordinate
- As can be seen from this diagram, the length AP is the y-coordinate of P. The x-coordinate of P you know is m, i.e. $x = m$.

HMSG: p 9

Q8. (c)

The area of the
rectangle, $A(m)$
is given by:

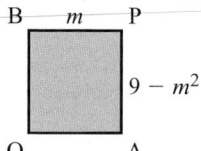

$A(m) = m \, (9 - m^2)$ ✓

$= 9m - m^3 \Rightarrow A'(m) = 9 - 3m^2$ ✓

For stationary value set $A'(m) = 0$ ✓

$\Rightarrow 9 - 3m^2 = 0 \Rightarrow 3m^2 = 9$

$\Rightarrow m^2 = 3$

So $m = \sqrt{3}, \; m \neq -\sqrt{3}$ since $m > 0$ ✓

m:

$A'(m) = 9 - 3m^2$:

Shape of graph:

nature: max ✓

So $m = \sqrt{3}$ gives a maximum value
for the area of rectangle $OAPB$. ✓

6 marks

Communication
- Area of a rectangle is given by length $\times$ breadth, and in this case the dimensions are m units $\times$ $(9 - m^2)$ units.

Differentiate
- The function $A(m)$ gives the area of the rectangle, whereas $A'(m)$ is the gradient function and determines the slope of the graph $y = A(m)$ showing the various areas as m changes.

Strategy
- Setting $A'(m)$ to zero determines which values of m give stationary points on the Area graph.

Solve
- Here there is a positive and a negative value of m. Only the positive value makes sense, as you are told that $0 < m < 3$.

Justify
- The 'nature' table is required to show that $m = \sqrt{3}$ does give a maximum value for the Area.

Communication
- A clear statement summarising your findings is needed for this final mark.

HMSG: pp 92–93

Q8. (d)

$A\left(\sqrt{3}\right) = \sqrt{3}\left(9 - \left(\sqrt{3}\right)^2\right)$

$= \sqrt{3}(9 - 3)$

$= \sqrt{3} \times 6$ ✓

So $6\sqrt{3}$ unit^2 is the maximum area.

1 mark

Calculation
- When you read the word 'exact' you know to steer clear of decimal approximations. Your answer will be an integer value, a fraction (rational number) or a surd (involving roots) or perhaps an expression involving π or e, ... but never approximate decimals.

HMSG: p 93

Answers to Exam C

Exam C: Paper 1

Q1. (a)

$$\vec{AB} = b - a = \begin{pmatrix} 7 \\ 14 \\ 21 \end{pmatrix} - \begin{pmatrix} 5 \\ 16 \\ 25 \end{pmatrix} = \begin{pmatrix} 2 \\ -2 \\ -4 \end{pmatrix} = 2\begin{pmatrix} 1 \\ -1 \\ -2 \end{pmatrix} \checkmark$$

$$\vec{BC} = c - b = \begin{pmatrix} 10 \\ 11 \\ 15 \end{pmatrix} - \begin{pmatrix} 7 \\ 14 \\ 21 \end{pmatrix} = \begin{pmatrix} 3 \\ -3 \\ -6 \end{pmatrix} = 3\begin{pmatrix} 1 \\ -1 \\ -2 \end{pmatrix} \checkmark$$

$$\Rightarrow \vec{BC} = \tfrac{3}{2}\vec{AB} \qquad\qquad \checkmark$$

So AB is parallel to BC. Also, since B is a shared point A, B and C are collinear. $\checkmark$

4 marks

Components
- To help with your calculation it is useful to rewrite $\vec{AB}$ in terms of position vectors:
$$\vec{AB} = b - a$$

Components
- Taking out the factors 2 and 3 is useful for the next part of the question.

Equation
- You should double check your equation on a pair of the components, e.g. the x-components:
$3 = \tfrac{3}{2} \times 2$. It is easy to mix it up and write $\tfrac{2}{3}$!

Statement
- To gain this mark it is essential that you mention 'parallel' and 'shared point'.

Q1. (b)

Here is a diagram:

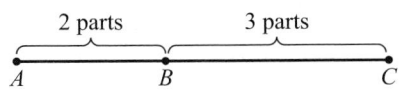

B divides AC in the ratio 2:3 $\checkmark$

1 mark

Ratio
- Always draw a sketch to help with the ratio and pay attention to the order of the letters:
AB (ratio 2:3) not BA (ratio 3:2)

HMSG: p 37

Q2. (a)
$$p(x) = f(g(x)) = f(\tfrac{1}{2}x + 1)$$
$$= 4(\tfrac{1}{2}x + 1) - 1 \qquad \checkmark$$
$$= 2x + 4 - 1 = 2x + 3 \qquad \checkmark$$

2 marks

Substitution
- f acts by multiplying by 4 then subtracting 1, in this case it acts like this on $\tfrac{1}{2}x + 1$

Simplificatin
- Don't leave brackets or unfinished calculations in a final answer expression.

Q2. (b)
$$h(p(x)) = h(2x + 3)$$
$$= \tfrac{1}{2}(2x + 3 - 3) \qquad \checkmark$$
$$= \tfrac{1}{2} \times 2x = x \qquad \checkmark$$

2 marks

Substitution
- h acts by subtracting 3 then halving the result, in this case it acts like this on $2x + 3$

Simplification
- Simplification must be completed for this mark.

Q2. (c)
h is the inverse of p (or p is the inverse of h). $\checkmark$

1 mark

Statement
- Put x into p then h sends you back to x again. h just undoes whatever p does: it is the inverse!

HMSG: p 10

Q3. The graph $y = -f(\frac{1}{2}x)$

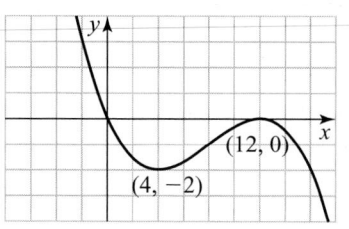

(12, 0)

(4, −2)

3 marks

x-axis scaling
- Think of $y = \sin(\frac{1}{2}x)$ which stretches out the $y = \sin x$ graph by a factor of 2 along the x-axis.

Flip in x-axis
- The appearance of $-1\times \ldots$ at the front indicates a reflection in the x-axis.

Annotation
- Make sure you show where all known points end up, e.g. (2, 2), (6, 0) and the origin.

HMSG: pp 14–15

Q4.
$$v = (\sin t)^{\frac{1}{2}}$$

$$\frac{dv}{dt} = \frac{1}{2}(\sin t)^{-\frac{1}{2}} \times \cos t$$

$$= \frac{\cos t}{2(\sin t)^{\frac{1}{2}}} = \frac{\cos t}{2\sqrt{\sin t}}$$

3 marks

Strategy & preparation
- For 'rate of change' you differentiate, so your preparation is to change the root to the power of $\frac{1}{2}$.

Differentiate a power
- This is for writing $\frac{1}{2}(\sin t)^{-\frac{1}{2}\ldots}$

Chain rule
- … and this is for remembering to differentiate $\sin t$ to get the factor $\times \cos t$

HMSG: pp 64–65

Q5.

$$2\begin{array}{|ccccc} 1 & -5 & a & b & -2 \\ & 2 & -6 & 2a-12 & 4a+2b-24 \\ \hline 1 & -3 & a-6 & 2a+b-12 & 4a+2b-26 \end{array}$$

Since $x - 2$ is a factor the remainder is zero

so $4a + 2b - 26 = 0 \Rightarrow 2a + b = 13$

$$-1\begin{array}{|ccccc} 1 & -5 & a & b & -2 \\ & -1 & 6 & -a-6 & a-b+6 \\ \hline 1 & -6 & a+6 & -a+b-6 & a-b+4 \end{array}$$

Division by $x + 1$ gives a remainder of 12
so $a - b + 4 = 12 \Rightarrow a - b = 8$

Now, solve: $\left.\begin{array}{l} 2a+b=13 \\ a-b=8 \end{array}\right\} \Rightarrow$ add: $3a = 21 \Rightarrow$ $a = 7$

Put $a = 7$ in $a - b = 8 \Rightarrow 7 - b = 8 \Rightarrow b = -1$

5 marks

Strategy
- Use of synthetic division leading to simultaneous equations gains this strategy mark

1st equation
- Simplifying the equation should leave the letters on the left and the constant on the right, in preparation for using simultaneous equations.
- Cancelling a factor of 2 from both sides helps keep the numbers smaller and more manageable.

2nd equation
- Be careful with the negative sign in calculations.

1st value
- Values should be clearly labeled to avoid any confusion.

2nd value
- Always check your values in both equations to verify they work!

HMSG: p 50–51, p 53

Q6.

$2x - 3y + 1 = 0 \Rightarrow 3y = 2x + 1 \Rightarrow y = \frac{2}{3}x + \frac{1}{3}$ ✓

$\Rightarrow m = \frac{2}{3} \Rightarrow m_\perp = -\frac{3}{2}$ ✓

The gradient is $-\frac{3}{2}$ and a point on line is $(-3, 2)$

The equation is $y - 2 = -\frac{3}{2}(x - (-3))$

$\Rightarrow 2y - 4 = -3(x + 3) \Rightarrow 2y + 3x = -5$ ✓

Gradient
- Rearrange to the form: $y = mx + c$ and state the value of m (the gradient)

Perpendicular gradient
- Use the result $m = \frac{a}{b} \Rightarrow m_\perp = -\frac{b}{a}$. In practical terms this means 'invert and change the sign'.

Equation
- Other forms are acceptable providing they have no brackets or uncompleted calculations,
  e.g. $2y + 3x + 5 = 0$ or $y = -\frac{3}{2}x - \frac{5}{2}$

HMSG: pp 77–79

Q7. (a)

$u_0 = 10$

$\Rightarrow u_1 = 0 \cdot 3\,u_0 + k = 0 \cdot 3 \times 10 + k = k + 3$ ✓

$\Rightarrow u_2 = 0 \cdot 3\,u_1 + k = 0 \cdot 3(k + 3) + k = 1 \cdot 3k + 0 \cdot 9$ ✓

So $1 \cdot 3k + 0 \cdot 9 = 7 \cdot 4$ ✓

$\Rightarrow 1 \cdot 3k = 6 \cdot 5 \Rightarrow k = \frac{6 \cdot 5}{1 \cdot 3} = 5$

1st term
- An expression involving k is required. You will not get a numerical value at this stage.

2nd term
- Be careful to multiply the whole of $k + 3$ by $0 \cdot 3$.

Value
- You now know u_2 is $1 \cdot 3k + 0 \cdot 9$ but you are also told it is $7 \cdot 4$. So they are equal and this gives an equation to solve to find k.

HMSG: pp 88–89

Q7. (b)

$u_{n+1} = 0 \cdot 3u_n + 5$ gives a sequence with a limit since the multiplier $0 \cdot 3$ lies between -1 and 1. ✓

1 mark

Statement
- $u_{n+1} = mu_n + c$ gives a limit whenever $-1 < m < 1$

Q7. (c)

$u_{n+1} = 0 \cdot 3u_n + 5$ gives a limit L

$\Rightarrow L = 0 \cdot 3L + 5 \Rightarrow L - 0 \cdot 3L = 5$ ✓

$\Rightarrow 0 \cdot 7L = 5 \Rightarrow L = \frac{5}{0 \cdot 7} = \frac{50}{7} = 7\frac{1}{7}$ ✓

2 marks

Strategy
- You may be used to using the formula:
  $L = \dfrac{c}{1 - m}$ for the sequence from $u_{n+1} = mu_n + c$

Calculation
- You have no access to a calculator in Paper 1 so you know an exact value is required.

HMSG: p 90

Q8.

$f'(x) = \sin 2x$

$\Rightarrow f(x) = \int \sin 2x\, dx = \frac{-\cos 2x}{2} + C$ ✓

Since $(\frac{\pi}{6}, -1)$ lies on the graph then $f(\frac{\pi}{6}) = -1$

$\Rightarrow \dfrac{-\cos 2 \times \frac{\pi}{6}}{2} + C = -1$ ✓

$\Rightarrow \dfrac{-\cos \frac{\pi}{3}}{2} + C = -1 \Rightarrow \dfrac{-\frac{1}{2}}{2} + C = -1$

$\Rightarrow \dfrac{-1}{4} + C = -1 \Rightarrow C = -1 + \frac{1}{4} = -\frac{3}{4}$

So $f(x) = -\frac{1}{2}\cos 2x - \frac{3}{4}$ ✓

Strategy
- From $f(x)$ to $f'(x)$ involves differentiation and from $f'(x)$ to $f(x)$ involves integration.
- Don't forget C, the constant is crucial to the solution of the question!

Equation and substitution
- If (a, b) lies on $y = f(x)$ then $b = f(a)$. In other words, the coordinates of the point satisfy the equation of the graph.

Formula
- You need to be able to work out exact values like $\cos \frac{\pi}{3}$ without a calculator.
- The equation asks for $f(x)$ not just for the value of C. Be careful you answer the question asked.

HMSG: p 24, p 71

3 marks

Q9. (a)

$y = x^3 - 3x^2 - 24x - 28$ ✓

$\Rightarrow \frac{dy}{dx} = 3x^2 - 6x - 24$ ✓

For stationary points set $\frac{dy}{dx} = 0$ ✓

$\Rightarrow 3x^2 - 6x - 24 = 0$

$\Rightarrow 3(x^2 - 2x - 8) = 0$

$\Rightarrow 3(x + 2)(x - 4) = 0$

$\Rightarrow x + 2 = 0$ or $x - 4 = 0$

$\Rightarrow x = -2$ or $x = 4$. ✓

$x:$

$\frac{dy}{dx} = 3(x + 2)(x - 4):$

Shape of graph:

nature: max min ✓

when $x = -2$

$y = (-2)^3 - 3 \times (-2)^2 - 24 \times (-2) - 28$

$= -8 - 12 + 48 - 28 = 0$

so $(-2, 0)$ is a maximum stationary point

when $x = 4$

$y = 4^3 - 3 \times 4^2 - 24 \times 4 - 28$

$= 64 - 48 - 96 - 28 = -108$ ✓

so $(4, -108)$ is a minimum stationary point ✓

7 marks

Q9. (b)

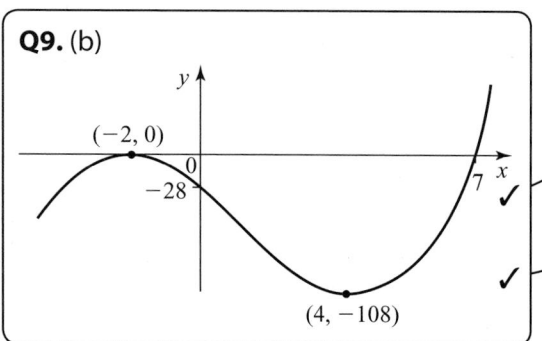

$(-2, 0)$

-28

7 ✓ ✓

$(4, -108)$

2 marks

Q10. (a)

From the diagram:

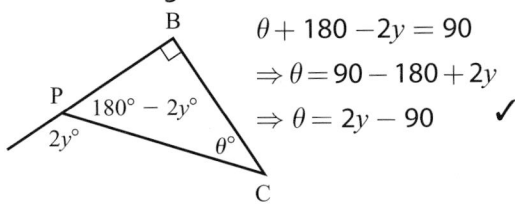

$\theta + 180 - 2y = 90$

$\Rightarrow \theta = 90 - 180 + 2y$

$\Rightarrow \theta = 2y - 90$ ✓

Angle
- 'Find θ in terms of y.' means you are aiming for:
  $\theta = $ (an expression involving y only) So θ cannot appear on the 'right-hand side'.
- You are using: the angle sum in a triangle is $180°$.

1 mark

Q10. (b)

$\sin\theta° = \sin(2y-90)°$
$= \sin 2y° \cos 90° - \cos 2y° \sin 90°$ ✓
$= \sin 2y° \times 0 - \cos 2y° \times 1$ ✓
$= -\cos 2y° = -(2\cos^2 y° - 1)$ ✓
$= -2\cos^2 y° + 1 = 1 - 2\cos^2 y°$

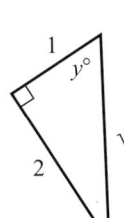

now $\cos y° = \dfrac{1}{\sqrt{5}}$ ✓

So, $\sin\theta° = 1 - 2 \times \left(\dfrac{1}{\sqrt{5}}\right)^2$ ✓

$= 1 - 2 \times \dfrac{1}{5} = 1 - \dfrac{2}{5} = \dfrac{3}{5}$ ✓

6 marks

Expansion
- The formula:
  $\sin(A \pm B) = \sin a \cos B \pm \cos a \sin B$ is given to you in your exam.

Simplify
- The values
  $\cos 90° = 0$
  and $\sin 90° = 1$ are known from the graphs.

$\cos 90° = 0$

$\sin 90° = 1$

Remember, no calculators in Paper 1.

Strategy
- Use of the 'Double angle' formula.

Calculation
- The exact value of $\cos y°$ is obtained using Pythagoras' Theorem in $\triangle$ APD and then using SOHCAHTOA.

Value
- Substitution of $\frac{1}{\sqrt{5}}$ for $\cos y°$ in the expression.

Simplify
- Notice: $\left(\frac{1}{\sqrt{5}}\right)^2 = \frac{1}{\sqrt{5}} \times \frac{1}{\sqrt{5}} = \frac{1 \times 1}{\sqrt{5} \times \sqrt{5}} = \frac{1}{5}$

HMSG: pp 29–31

Q11. $\log_{\sqrt{2}} x - \log_{\sqrt{2}} 2 = 2$

$\Rightarrow \log_{\sqrt{2}} \dfrac{x}{2} = 2$ ✓ ✓

$\Rightarrow \dfrac{x}{2} = \left(\sqrt{2}\right)^2 = 2$ ✓

$\Rightarrow x = 4.$ ✓

4 marks

Log Law
- The 'law' used is $\log_b m - \log_b n = \log_b \dfrac{m}{n}$

Exponential form
- Rewrite $\log_b a = c$ as $a = b^c$.

Start Solution
- $\left(\sqrt{2}\right)^2 = \sqrt{2} \times \sqrt{2} = 2$

Finish Solutions
- $\dfrac{x}{2} = 2 \Rightarrow \dfrac{x}{2} \times 2 = 2 \times 2 \Rightarrow x = 4$

HMSG: p 20

Q12. (a)

$a = \frac{3}{2}$ ✓

$b = -\left(\frac{\pi}{2} + \frac{\pi}{3}\right) = -\left(\frac{3\pi}{6} + \frac{2\pi}{6}\right) = -\frac{5\pi}{6}$ ✓

$c = \frac{1}{2}$ ✓

3 marks

Value of a
- a is the amplitude. This is half the difference between the maximum and minimum values:
  $a = \frac{1}{2}(2 - (-1)) = \frac{1}{2} \times 3$

Value of b
- The 'normal' sine graph is shifted right so b will have a negative value.
- Minimum normally at $x = -\frac{\pi}{2}$ hence $\frac{\pi}{2} + \frac{\pi}{3}$ to the right for the new minimum

Value of c
- Halfway between min $= -1$ and max $= 2$. The 'normal' sine graph is shifted up $\frac{1}{2}$ unit.

HMSG: pp 26–27

Q12. (b)

The graph is $y = \frac{3}{2}\sin(x - \frac{5\pi}{6}) + \frac{1}{2}$
For the y-intercept set $x = 0$

$\Rightarrow y = \frac{3}{2}\sin(0 - \frac{5\pi}{6}) + \frac{1}{2} = y$ ✓

$\quad = \frac{3}{2}\sin(-\frac{5\pi}{6}) + \frac{1}{2}$

$\quad = \frac{3}{2} \times (-\frac{1}{2}) + \frac{1}{2} = -\frac{3}{4} + \frac{1}{2} = -\frac{1}{4}$ ✓

So $A(0, -\frac{1}{4})$

Strategy
- You need to construct the function formula from the values obtained in part **a**, then let $x = 0$ to find the y-intercept.

Coordinates
- $-\frac{5\pi}{6}$ is a 3rd quadrant angle (clockwise rotation from the x-axis) and so $\sin(-\frac{5\pi}{6})$ is negative with 1st quadrant angle $\frac{\pi}{6}$ so equals $-\sin\frac{\pi}{6}$
- Make sure you give the coordinates and not just the value of the y-coordinate

HMSG: p 24

Q12. (c)

For the points of intersection, put $y = \frac{5}{4}$ into the formula for the graph
$y = \frac{3}{2}\sin(x - \frac{5\pi}{6}) + \frac{1}{2}$

$\Rightarrow \frac{3}{2}\sin(x - \frac{5\pi}{6}) + \frac{1}{2} = \frac{5}{4}$

$\Rightarrow 6\sin(x - \frac{5\pi}{6}) + 2 = 5$

$\Rightarrow 6\sin(x - \frac{5\pi}{6}) = 3$

$\Rightarrow \sin(x - \frac{5\pi}{6}) = \frac{3}{6} = \frac{1}{2}$ ✓

So $x - \frac{5\pi}{6}$ is a 1st or 2nd quadrant angle (positive)

So $x - \frac{5\pi}{6} = \frac{\pi}{6}$ or $x - \frac{5\pi}{6} = \frac{5\pi}{6}$

$\Rightarrow x = \frac{\pi}{6} + \frac{5\pi}{6}$ or $x = \frac{5\pi}{6} + \frac{5\pi}{6}$

$\Rightarrow x = \frac{6\pi}{6} = \pi$ or $x = \frac{10\pi}{6} = \frac{5\pi}{3}$ ✓

From the diagram, $x = \pi$ gives intersection point B

Simplified equation
- Get rid of fractions initially by multiplying through by 4.
- Simplification leads to the form: sin(angle) = number

Coordinate value
- You should make sure that your answer seems reasonable in the context of the given graph:
- Is $\frac{\pi}{3}$ on the x-axis $\frac{1}{3}$ of the way towards B from the origin? Yes! It looks good.

HMSG: p 54, p 56

Exam C: Paper 2

Q1.

$\sin 2x - \sqrt{3}\,\sin x = 0$ ✓

$\Rightarrow 2\sin x \cos x - \sqrt{3}\,\sin x = 0$

$\Rightarrow \sin x\,(2\cos x - \sqrt{3}) = 0$ ✓

$\Rightarrow \sin x = 0$ or $\cos x = \dfrac{\sqrt{3}}{2}$ ✓

For $\sin x = 0$
$x = 0, \pi, 2\pi$

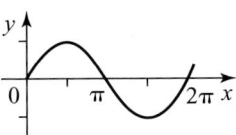

For $\cos x = \dfrac{\sqrt{3}}{2}$
x is in 1st or 4th quadrants
$\quad 1^{\text{st}}$ quadrant angle is $\dfrac{\pi}{6}$

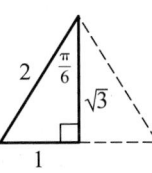

So

$x = \dfrac{\pi}{6}$ or $2\pi - \dfrac{\pi}{6} = \dfrac{12\pi}{6} - \dfrac{\pi}{6} = \dfrac{11\pi}{6}$ ✓

Solutions are:

$$x = 0, \frac{\pi}{6}, \pi, \frac{11\pi}{6}, 2\pi$$

5 marks

Strategy
- Using the 'Double Angle Formula' allows you to factorise the expression. The formula $\sin 2A = 2\sin A \cos A$ is given to you in the exam.

Factorise
- The common factor is $\sin x$

Solve
- Notice $2\cos x - \sqrt{3} = 0$
  $$\Rightarrow 2\cos x = \sqrt{3} \Rightarrow \cos x = \tfrac{\sqrt{3}}{2}$$

Angles
- For $\sin x$ or $\cos x$ equal to values $-1, 0$ or 1, you should use the $y = \sin x$ or $y = \cos x$ graph to determine the angles.

- Take care that the angles you give as solutions are allowed. In this case, the interval in which the angles lie is $0 \le x \le 2\pi$.

- For quadrant use:
  ✓ shows cosine positive

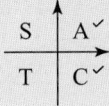

HMSG: pp 54–55

Q2. (a)

$3\sin x° - \cos x° = k\sin(x-a)°$

$\Rightarrow 3\sin x° - \cos x°$

$= k\sin x°\cos a° - k\cos x°\sin a°$ ✓

Now equate the coefficients of

$\sin x°$ and $\cos x°$

$\left.\begin{array}{l} k\cos a° = 3 \\ \\ k\sin a° = 1 \end{array}\right\}$ since both $\sin a°$ and
$\cos a°$ are positive, $a°$ is
in the 1st quadrant. ✓

Divide:

$\dfrac{k\sin a°}{k\cos a°} = \dfrac{1}{3} \Rightarrow \tan a° = \dfrac{1}{3}$

so $a° = \tan^{-1}\left(\dfrac{1}{3}\right) \doteq 18\cdot4°$ ✓

(to 1 decimal place)

Square and add:

$(k\cos a°)^2 + (k\sin a°)^2 = 3^2 + 1^2$

$\Rightarrow k^2\cos^2 a° + k^2\sin^2 a° = 9 + 1$

$\Rightarrow k^2(\cos^2 a° + \sin^2 a°) = 10$

$\Rightarrow k^2 \times 1 = 10 \Rightarrow k^2 = 10$

$\Rightarrow k = \sqrt{10} \ (k > 0)$ ✓

So $3\sin x° - \cos x°$

$= \sqrt{10}\sin(x - 18\cdot4)°$

(correct to 1 decimal place)

4 marks

Strategy

- Your first step is to expand $k\sin(x-a)°$.
- On your formulae sheet is:

  $\sin(A \pm B) = \sin A\cos B \pm \cos A\sin B$.

  This is the expansion you should use.

- Notice that there is an implied step in the working:

  $k\sin(x-a)° = k(\sin x°\cos a° - \cos x°\sin a°)$
  $\qquad\qquad = k\sin x°\cos a° - k\cos x°\sin a°$

- If this expansion does not appear in your working, you will not be awarded this mark.

Coefficients

- The method is:

  $\textcircled{3}\ \sin x° \qquad -\textcircled{1}\ \cos x°$

  $\underbrace{k\sin x°\cos a°} - \underbrace{k\cos x°\sin a°}$

  so $k\cos a° = 3$ and $k\sin a° = 1$.

Angle

- A common mistake is to divide in the wrong order. The result you are using is $\dfrac{\sin a°}{\cos a°} = \tan a°$

  In this case, since $k\sin a°$ is on the 'top' of the fraction but comes from the 'bottom' equation so the fraction is $\frac{1}{3}$ not $\frac{3}{1}$.

Amplitude

- You should try to understand the method for calculating k. This will help you deal with more unusual questions that sometimes arise – these require understanding, not the blind application of a formula.

HMSG: pp 32–33

Q2. (b)

$3 \sin x° - \cos x° = 1$

$\Rightarrow \sqrt{10} \sin(x - 18\cdot4...)° = 1$ ✓

$\Rightarrow \sin(x - 18\cdot4...)° = \dfrac{1}{\sqrt{10}}$

$\Rightarrow x° - 18\cdot4...° = \sin^{-1}\left(\dfrac{1}{\sqrt{10}}\right) = 18\cdot4...°$ ✓

So $x = 18\cdot43... + 18\cdot43...$

$\quad = 36\cdot86...$

$\Rightarrow x \doteqdot 36\cdot9$ (to decimal place) ✓

$\quad (0 \leq x \leq 90)$

Strategy
- Using your result from part (a) you can rewrite the given equation in a form that is easier to solve.

Solve for $(x-a)°$
- Your aim is to move to an equation of the form $\sin(\text{angle}) = \text{number}$.
- Since the interval of allowable values of x is $0 \leq x \leq 90$ only, the 1st quadrant value is considered in this particular example.

Solve for x
- $x - 18\cdot4 = 18\cdot4 \Rightarrow x = 18\cdot4 + 18\cdot4$

HMSG: p 54

3 marks

Q3.

Use $\cos\theta = \dfrac{\boldsymbol{u}\cdot\boldsymbol{w}}{|\boldsymbol{u}||\boldsymbol{w}|}$

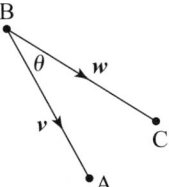

When $\boldsymbol{v} = \begin{pmatrix} -2 \\ 3 \\ 5 \end{pmatrix}$ and $\boldsymbol{w} = \begin{pmatrix} 1 \\ -1 \\ 3 \end{pmatrix}$ ✓

$\boldsymbol{u}\cdot\boldsymbol{w} = \begin{pmatrix} -2 \\ 3 \\ 5 \end{pmatrix} \cdot \begin{pmatrix} 1 \\ -1 \\ 3 \end{pmatrix}$

$\quad = -2 \times 1 + 3 \times (-1) + 5 \times 3 = 10$ ✓

$|\boldsymbol{v}| = \sqrt{(-2)^2 + 3^2 + 5^2} = \sqrt{4 + 9 + 25}$

$\quad = \sqrt{38}$ ✓

$|\boldsymbol{w}| = \sqrt{1^2 + (-1)^2 + 3^2} = \sqrt{1 + 1 + 9}$

$\quad = \sqrt{11}$ ✓

So $\cos\theta = \dfrac{10}{\sqrt{38}\ \sqrt{11}}$

$\Rightarrow \theta = \cos^{-1}\left(\dfrac{10}{\sqrt{38}\ \sqrt{11}}\right)$ ✓

$\Rightarrow \theta = 60\cdot71 \ldots \doteqdot 60\cdot7$ (to 1 dec. place)

Strategy
- The formula given to you in the exam is: '$\boldsymbol{a}\cdot\boldsymbol{b} = |\boldsymbol{a}||\boldsymbol{b}|\cos\theta$, where θ is the angle between $\boldsymbol{a}$ and $\boldsymbol{b}$ so rearrange into the form used in this question:
$\cos\theta = \dfrac{\boldsymbol{a}\cdot\boldsymbol{b}}{|\boldsymbol{a}||\boldsymbol{b}|}$
As is the case for all formulae: LEARN THEM!

Dot product
- There is a check you can do with the 'dot product'
- If $\boldsymbol{a}\cdot\boldsymbol{b} > 0$ the angle is acute $(0° < \theta° < 90°)$
- If $\boldsymbol{a}\cdot\boldsymbol{b} < 0$ then the angle is obtuse $(90° < \theta° < 180°)$

Magnitudes
- You will be awarded 1 mark for each correct answer.

Angle
- Many mistakes are made in this calculation. Use:

$\boxed{\text{INV}}\ \boxed{\cos}\ \boxed{(}\ \boxed{1}\boxed{0}\ \boxed{\div}\ \boxed{(}\ \boxed{\sqrt{}}\ \boxed{3}\boxed{8}\ \boxed{\times}\ \boxed{\sqrt{}}\ \boxed{1}\boxed{1}\ \boxed{)}\ \boxed{)}$

as the keying sequence followed by $\boxed{\text{EXE}}$ (true for most calculators).

HMSG: p 38

5 marks

Q4.

$x^2 - 2x + c^2 + 2 = 0$ ✓

Discriminant

$= (-2)^2 - 4 \times 1 \times (c^2 + 2)$ ✓

$= 4 - 4(c^2 + 2) = 4 - 4c^2 - 8$

$= -4c^2 - 4 = -4(c^2 + 1)$ ✓

Since $c^2 + 1 > 0$ then $-4(c^2 + 1) < 0$

So, for all values of c the Discriminant < 0 and the equation has no Real roots. ✓

4 marks

Strategy
- The positive/zero/negative nature of the discriminant of this quadratic equation has to be determined.

Substitution
- Comparing $ax^2 + bx + c' = 0$

  with $1x^2 - 2x + (c^2 + 2) = 0$

  gives $a = 1$, $b = -2$ and $c' = c^2 + 2$ (different cs!)

  So $b^2 - 4ac' = (-2)^2 - 4 \times 1 \times (c^2 + 2)$

Simplify
- Be careful with negatives:
  e.g. $-4(c^2 + 2) = -4c^2 - 8$

Proof
- $-4 \times (c^2 + 1)$ is negative $\times$ positive $=$ negative

  $c^2 + 1$ is always positive since c^2 is always positive or zero.

HMSG: p 46, p 48

Q5. (a)

$$y = \frac{1}{3}x^3 - 2x^2$$ ✓

$$\Rightarrow \frac{dy}{dx} = x^2 - 4x$$ ✓

The tangent has gradient -4

so $\dfrac{dy}{dx} = -4$ ✓

$\Rightarrow x^2 - 4x = -4$

$\Rightarrow x^2 - 4x + 4 = 0$ ✓

$\Rightarrow (x - 2)(x - 2) = 0$

$\Rightarrow x = 2$ which is the required
  x-coordinate. ✓

5 marks

Strategy
- Knowing to differentiate will earn you this mark.

Differentiate
- Notice the coefficients: $3 \times \dfrac{1}{3} = 1$
  and $2 \times (-2) = -4$.

Strategy
- This 2nd strategy mark is for setting the gradient equal to -4.

Solve
- 2 marks are available: for starting the process and then completing it. You should recognise a quadratic equation and write it in 'standard form', i.e. $x^2 - 4x + 4 = 0$.

HMSG: pp 58–59

Q5. (b)

When $x = 2$

$$y = \frac{1}{3} \times 2^3 - 2 \times 2^2 = \frac{8}{3} - 8 = -\frac{16}{3} \quad \checkmark$$

A point on the tangent is

$$\left(2, -\frac{16}{3}\right)$$

and the gradient $= -4$
So the equation is:

$$y - \left(-\frac{16}{3}\right) = -4(x - 2)$$

$$\Rightarrow y + \frac{16}{3} = -4x + 8$$

$$\Rightarrow 3y + 16 = -12x + 24$$

$$\Rightarrow 3y + 12x = 8 \quad \checkmark$$

2 marks

Calculation
- To find the equation of the tangent you will need to know the coordinates of a point on that tangent. All you know so far is the x-coordinate ($x = 2$), and so calculation of the y-coordinate is essential.

Equation
- Here you are using $y - b = m(x - a)$ where $m = -4$ and the point (a, b) is $\left(2, -\frac{16}{3}\right)$
- Do not use decimal approximations in coordinate work — you will lose marks if you do, e.g. $-5 \cdot 3$ should not be used for $-\frac{16}{3}$

HMSG: p 60

Q6. (a)

$$C_t = C_o e^{-\frac{t}{4}}$$

In this case $C_t = 3 \cdot 5$ when $t = 3$

$$\Rightarrow 3 \cdot 5 = C_o e^{-\frac{3}{4}} \Rightarrow 3 \cdot 5 = \frac{C_o}{e^{\frac{3}{4}}} \quad \checkmark$$

$$\Rightarrow C_o = 3 \cdot 5 e^{\frac{3}{4}} = 7 \cdot 409\ldots \quad \checkmark$$

The concentration just after administration was $7 \cdot 4$ mg/ml (to 1 decimal place). $\quad \checkmark$

3 marks

Strategy
- In a question like this, you should know that you will have to pick out values for the letters in the formula and do a substitution. Sometimes it helps to label the formula:

$$C_t = C_o e^{-\frac{t}{4}}$$

Concentration after t hours · concentration at the start · the time elapsed (t hours)

Change of Subject.
- C_t is the subject of the given formula. You are asked to find C_0 — the concentration initially. This means you aim for $C_0 = $ (a number)

Calculation
- On the calculator use:

$$\boxed{3}\,\boxed{\bullet}\,\boxed{5}\,\boxed{\times}\,\boxed{e^x}\,\boxed{(}\,\boxed{3}\,\boxed{\div}\,\boxed{4}\,\boxed{)}\,\boxed{\text{EXE}}$$

HMSG: p 22

Q6. (b)

Required to find t so that:

$C_t = \frac{1}{2}C_o$

$\Rightarrow \frac{1}{2}C_o = C_o e^{-\frac{t}{4}}$ ✓

✓

$\Rightarrow \frac{1}{2} = e^{-\frac{t}{4}}$ ✓

$\Rightarrow \log_e \frac{1}{2} = -\frac{t}{4}$

$\Rightarrow t = -4\log_e \frac{1}{2} = 2.772...$

This is 2 hours and 0.772... $\times$ 60 min. It takes 2 hours 46 minutes (to the nearest minute). ✓

4 marks

Interpretation
- You are being asked to calculate t for C_t (the concentration after t hours) to equal half of the initial concentration, i.e. $\frac{1}{2}C_o$

1st step to solving
- Both sides of the equation can be divided by C_0. This creates an exponential equation with only the variable t.

Log Statement
- You use this conversion:

$c = b^a \leftrightarrow \log_b c = a$

In this case $a = -\frac{t}{4}$, $b = e$ and $c = \frac{1}{2}$

Calculation
- The $\boxed{\ln}$ key calculates '$\log_e$'.

HMSG: p 22

Q7. (a)

$x^2 + y^2 - 4x - 6y + 8 = 0$

Centre: $C_1(2, 3)$ ✓

For $C_1(2, 3)$ and $A(1, 5)$ ✓

$m_{CA} = \frac{5-3}{1-2} = \frac{2}{-1} = -2$ ✓

$\Rightarrow m_\perp = \frac{1}{2}$

Point on the tangent is $A(1, 5)$ and the gradient $= \frac{1}{2}$

So the equation is:

$y - 5 = \frac{1}{2}(x - 1)$

$\Rightarrow 2y - 10 = x - 1$

$\Rightarrow 2y - x = 9$ ✓

4 marks

Centre
- This uses the result:

circle: $x^2 + y^2 + 2gx + 2fy + c = 0$

Centre: $(-\dot{g}, \dot{-}f)$

The process involves halving and changing signs of the coefficients of x and y to get the coordinates of the centre of the circle.

Gradient of radius
- The gradient formula is:

$P(x_1, y_1), Q(x_2, y_2) \quad m_{PQ} = \frac{y_2 - y_1}{x_2 - x_1}$

Strategy
- The tangent is perpendicular to the radius to the point of contact, i.e. $C_1 P$
- Perpendicular gradients are obtained by using

$m = \frac{a}{b} \Rightarrow m_\perp = -\frac{b}{a}$ In this case -2 is thought of as $-\frac{2}{1}$

Equation
- Use $y - b = m(x - a)$. In this case, you use $m = \frac{1}{2}$ with (a, b) being the point $A(1, 5)$.

HMSG: pp 76–79, p 82, p 84

Q7. (b)

For intersection of line and circle solve:

$$2y - x = 9$$
$$x^2 + y^2 + 2x + 2y - 18 = 0$$ ✓

Substitute $x = 2y - 9$ in circle equation: ✓

$(2y - 9)^2 + y^2 + 2(2y - 9) + 2y - 18 = 0$

$\Rightarrow 4y^2 - 36y + 81 + y^2 + 4y - 18$
$\quad + 2y - 18 = 0$

$\Rightarrow 5y^2 - 30y + 45 = 0$ ✓

$\Rightarrow 5(y^2 - 6y + 9) = 0$

$\Rightarrow 5(y - 3)(y - 3) = 0$

$\Rightarrow y = 3$ ✓

and since there is only one solution, the line is a tangent to the circle. ✓

5 marks

Rearrangement
- Change $2y - x = 9$ to $x = 2y - 9$ ready for substitution.

Strategy
- The strategy mark here is awarded for substitution of the line equation into the circle equation.

'Standard form'
- You should recognise a quadratic equation and therefore write it in the standard order, namely:
  $5y^2 - 30y + 45 = 0$

Solve
- It is always easier to factorise quadratic expressions if any common factor is removed first. In this case, 5 is the common factor.

Justify
- How do you prove a line is tangent to a circle? You find the points of intersection. If the line is a tangent there will be only one point. You write a clear statement to this effect to gain this 'communication' work.

HMSG: p 84

Q7. (c)

When $y = 3$

$x = 2 \times 3 - 9 = -3$ ✓

So the point of contact is $B(-3, 3)$

For $A(1, 5)$ and $B(-3, 3)$

$AB = \sqrt{(1 - (-3))^2 + (5 - 3)^2}$

$\quad = \sqrt{4^2 + 2^2} = \sqrt{20} = 2\sqrt{5}$ ✓

2 marks

Other coordinate
- Having determined the value of the y-coordinate in part (b), you now have to substitute this back (use the line equation) to find the x-coordinate.

Distance
- You use the distance formula: $P(x_1, y_1)$, $Q(x_2, y_2)$
  $PQ = \sqrt{(x_1 - x_2)^2 + (y_1 - y_2)^2}$

HMSG: pp 80–84

Q8. ✓

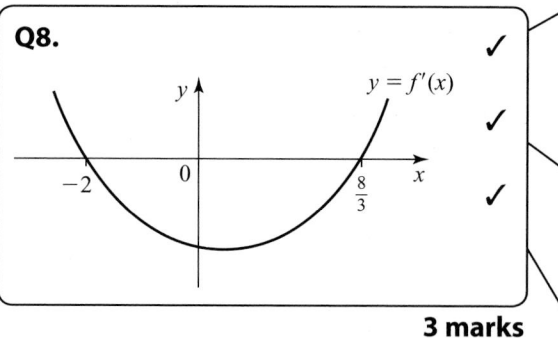

3 marks

Stationary points
- Any stationary point on the graph $y = f(x)$ has a zero value for the gradient. So, on $y = f'(x)$, the graph that shows the gradient values, an x-axis intercept (zero value) indicates a stationary point on the original graph. In this case, your sketch should cross the x-axis at $(-2, 0)$ and $(\frac{8}{3}, 0)$.

Correct relationship
- Between -2 and $\frac{8}{3}$ the original graph goes downhill:

  $\Rightarrow$ gradient is negative $\Rightarrow y = f'(x)$ graph is below the x-axis

- Less than -2 and greater than $\frac{8}{3}$ original graph goes uphill: $\Rightarrow$ gradient positive $\Rightarrow y = f'(x)$ graph is above the x-axis.

Shape
- Differentiating a cubic produces a quadratic $\Rightarrow$ your graph should show a parabola.

HMSG: p 63

Q9. (a)

Volume $= x \times x \times h = x^2 h$ ✓

so $x^2 h = 62\frac{1}{2} \Rightarrow h = \dfrac{62\frac{1}{2}}{x^2} = \dfrac{125}{2x^2}$

base area $= x^2$

side area $= xh$

$= x \times \dfrac{125}{2x^2} = \dfrac{125}{2x}$ ✓

total area $= 4 \times$ side area $+$ base area

$\Rightarrow A(x) = 4 \times \dfrac{125}{2x} + x^2 = \dfrac{250}{x} + x^2$ ✓

3 marks

Strategy
- Volume $=$ length $\times$ breadth $\times$ height.

Strategy
- Finding h in terms of x $\left(h = \frac{125}{2x^2}\right)$ and then substituting in the area expressions is the key to solving this question.

Proof
- Make sure all steps are shown. The answer is given, so the examiner needs to see each step of your reasoning.

Q9. (b)

$A(x) = 250x^{-1} + x^2$ ✓

$\Rightarrow A'(x) = -250x^{-2} + 2x$

$\quad = -\dfrac{250}{x^2} + 2x$ ✓

For stationary values set

$A'(x) = 0$ ✓

$\Rightarrow -\dfrac{250}{x^2} + 2x = 0 \;(\times x^2)$

$\Rightarrow -250 + 2x^3 = 0 \Rightarrow 2x^3 = 250$

$\Rightarrow x^3 = 125 \Rightarrow x = 5$ ✓

x:

$A'(x) = -\dfrac{250}{x^2} + 2x$:

Shape of graph:

nature: min ✓

So $x = 5$ gives a minimum value for the area.

5 marks

Preparation

- You should know that the term $\frac{250}{x}$ cannot be directly differentiated. Changing the expression to $250x^{-1}$ fits into the following rule:
  $f(x) = ax^n \Rightarrow f'(x) = nax^{n-1}$
  In this case $a = 250$ and $n = -1$

Differentiation

- It is useful for subsequent work to change negative indices to positive. In this case $-250x^{-2} = -\frac{250}{x^2}$

Strategy

- '$A'(x) = 0$' has to be stated to gain this mark.

Solve

- You should multiply both sides by x^2. The aim is to rid the equation of fractions.

- Cubing a positive number gives a positive, but cubing a negative number gives a negative, so $x^3 = 125$ has only one positive solution. This is not like squaring: $x^2 = 25$ gives $x = 5$ or $x = -5$.

Justify

- 1 mark is allocated for the 'nature table'.

HMSG: p 93

Q10.

First, find the points of intersection of $y = 5$ with the curve.

Solve:

$$\left.\begin{array}{l} y = 5 \\ y = x^2 - 6x + 10 \end{array}\right\} \quad x^2 - 6x + 10 = 5 \quad \checkmark$$

$$\Rightarrow x^2 - 6x + 5 = 0$$

$$\Rightarrow (x-1)(x-5) = 0 \quad \checkmark$$

$$\Rightarrow x = 1 \text{ or } x = 5$$

Diagram:

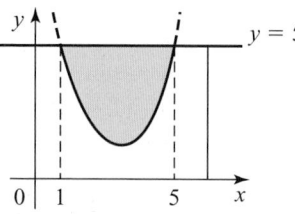

$\checkmark$

The shaded area in this diagram is given by:

$$\int_1^5 5 - (x^2 - 6x + 10)\, dx \quad \checkmark$$
$$\checkmark$$
$$= \int_1^5 5 - x^2 + 6x - 10\, dx$$
$$\checkmark$$
$$= \int_1^5 -x^2 + 6x - 5\, dx$$
$$= \left[-\frac{x^3}{3} + 3x^2 - 5x \right]_1^5$$
$$= \left(-\frac{5^3}{3} + 3 \times 5^2 - 5 \times 5 \right)$$
$$- \left(-\frac{1^3}{3} + 3 \times 1^2 - 5 \times 1 \right)$$
$$= -\frac{125}{3} + 75 - 25 + \frac{1}{3} - 3 + 5 \quad \checkmark$$
$$= 52 - \frac{124}{3} = \frac{156}{3} - \frac{124}{3} = \frac{32}{3}\ \text{cm}^2$$

So, area of plate $= 5 \times 6 \quad -\frac{32}{3}$

(area of rectangle)

$$= 30 - \frac{32}{3} = \frac{90}{3} - \frac{32}{3} = \frac{58}{3}$$

$\checkmark$

$$= 19\frac{1}{3}\ \text{cm}^2$$

8 marks

Strategy
- The strategy is to find the intersection points of line $y = 5$ and parabola $y = x^2 - 6x + 10$, as the x-values of these points will subsequently be crucial in setting up integrals to find areas.

Solve
- Always rearrange a quadratic equation into 'standard form', in this case $x^2 - 6x + 5 = 0$ so that the factorisation allows this process: (factor1)(factor2) $= 0 \Rightarrow$ factor 1 $= 0$ or factor 2 $= 0$

Strategy
- This strategy mark is for how you will split areas up. In the solution given the approach is:

  (rectangle area) $-$ (parabolic area)

  An alternative approach would have been to add together three areas to calculate the plate area as shown in this diagram.

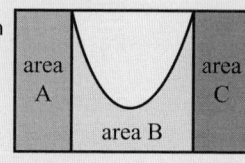

area A + area B + area C

Strategy
- Did you know to integrate to find this area?

Limits
- Left to right on the diagram (1 and 5). Bottom to top on the integral (1 and 5).

Integration
- You should always 'simplify' first before you integrate. In this case, don't integrate $5 - (x^2 - 6x + 10)$, simplify first to $-x^2 + 6x - 5$ and then integrate.

Evaluate
- Even with a calculator available these calculations are very difficult to get correct. You should take great care and double-check your working always.

Area
- Finally, make sure you answer the question. $\frac{32}{3}$ cm^2 is not the required answer! Your strategy was to subtract the parabolic area from the rectangle.

HMSG: p 94–95

Q11.

$f'(x) = x^2(x - 1) = x^3 - x^2$ ✓

$\Rightarrow f(x) = \int x^3 - x^2 \ dx$ ✓

✓

$= \dfrac{x^4}{4} - \dfrac{x^3}{3} + c$

Now $f(2) = \dfrac{1}{3}$ ✓

$\Rightarrow \dfrac{2^4}{4} - \dfrac{2^3}{3} + c = \dfrac{1}{3}$

$\Rightarrow 4 - \dfrac{8}{3} + c = \dfrac{1}{3}$ ✓

$\Rightarrow c = \dfrac{1}{3} + \dfrac{8}{3} - 4 = 3 - 4 = -1$

So $f(x) = \dfrac{1}{4}x^4 - \dfrac{1}{3}x^3 - 1$

5 marks

Strategy
- In questions where you are given $f'(x)$ (or $\dfrac{dy}{dx}$) and asked to find $f(x)$ (or y) then you need to use integration:

$$f(x) \xrightarrow{\ \text{differentiation}\ } f'(x)$$
$$f(x) \xleftarrow{\ \text{integration}\ } f'(x)$$

Preparation
- Multiplying out brackets is essential before you integrate.

Integrate
- You are using $\displaystyle\int x^n \, dx = \dfrac{x^{n+1}}{n+1} + c$

Substitution
- The 'constant of integration' c is crucial in this question. $f(2) = \frac{1}{3}$ means when $x = 2$ the formula gives $\frac{1}{3}$. c can therefore be found.

Calculation
- Exact values — so practice your fraction work!

HMSG: p 68

Notes

Notes

Notes

Notes

Notes